2016 SQA Past Papers With Answers

National 5

BIOLOGY

SQA National 5 BIOLOGY

2014, 2015 & 2016 Exams

This book contains the official SQA 2014, 2015 and 2016 Exams for National 5 Biology, with associated SQA-approved answers modified from the official marking instructions that accompany the paper.

In addition the book contains study skills advice. This has been specially commissioned by Hodder Gibson, and has been written by experienced senior teachers and examiners in line with the new National 5 syllabus and assessment outlines. This is not SQA material but has been devised to provide further guidance for National 5 examinations.

Hodder Gibson is grateful to the copyright holders, as credited on the final page of the Answer Section, for permission to use their material. Every effort has been made to trace the copyright holders and to obtain their permission for the use of copyright material. Hodder Gibson will be happy to receive information allowing us to rectify any error or omission in future editions.

Hachette UK's policy is to use papers that are natural, renewable and recyclable products and made from wood grown in sustainable forests. The logging and manufacturing processes are expected to conform to the environmental regulations of the country of origin.

Orders: please contact Bookpoint Ltd, 130 Park Drive, Milton Park, Abingdon, Oxon OX14 4SE. Telephone: (44) 01235 827720. Fax: (44) 01235 400454. Lines are open 9.00–5.00, Monday to Saturday, with a 24-hour message answering service. Visit our website at www.hoddereducation.co.uk. Hodder Gibson can be contacted direct on: Tel: 0141 333 4650; Fax: 0141 404 8188; email: hoddergibson@hodder.co.uk

This collection first published in 2016 by
Hodder Gibson, an imprint of Hodder Education,
An Hachette UK Company
211 St Vincent Street
Glasgow G2 5QY

Typeset by Aptara, Inc.

Printed in the UK

A catalogue record for this title is available from the British Library

ISBN: 978-1-4718-9102-1

3 2 1

2017 2016

Introduction

Study Skills – what you need to know to pass exams!

Pause for thought

Many students might skip quickly through a page like this. After all, we all know how to revise. Do you really though?

Think about this:

"IF YOU ALWAYS DO WHAT YOU ALWAYS DO, YOU WILL ALWAYS GET WHAT YOU HAVE ALWAYS GOT."

Do you like the grades you get? Do you want to do better? If you get full marks in your assessment, then that's great! Change nothing! This section is just to help you get that little bit better than you already are.

There are two main parts to the advice on offer here. The first part highlights fairly obvious things but which are also very important. The second part makes suggestions about revision that you might not have thought about but which WILL help you.

Part 1

DOH! It's so obvious but ...

Start revising in good time

Don't leave it until the last minute – this will make you panic.

Make a revision timetable that sets out work time AND play time.

Sleep and eat!

Obvious really, and very helpful. Avoid arguments or stressful things too – even games that wind you up. You need to be fit, awake and focused!

Know your place!

Make sure you know exactly **WHEN and WHERE** your exams are.

Know your enemy!

Make sure you know what to expect in the exam.

How is the paper structured?

How much time is there for each question?

What types of question are involved?

Which topics seem to come up time and time again?

Which topics are your strongest and which are your weakest?

Are all topics compulsory or are there choices?

Learn by DOING!

There is no substitute for past papers and practice papers – they are simply essential! Tackling this collection of papers and answers is exactly the right thing to be doing as your exams approach.

Part 2

People learn in different ways. Some like low light, some bright. Some like early morning, some like evening / night. Some prefer warm, some prefer cold. But everyone uses their BRAIN and the brain works when it is active. Passive learning – sitting gazing at notes – is the most INEFFICIENT way to learn anything. Below you will find tips and ideas for making your revision more effective and maybe even more enjoyable. What follows gets your brain active, and active learning works!

Activity 1 – Stop and review

Step 1

When you have done no more than 5 minutes of revision reading STOP!

Step 2

Write a heading in your own words which sums up the topic you have been revising.

Step 3

Write a summary of what you have revised in no more than two sentences. Don't fool yourself by saying, "I know it, but I cannot put it into words". That just means you don't know it well enough. If you cannot write your summary, revise that section again, knowing that you must write a summary at the end of it. Many of you will have notebooks full of blue/black ink writing. Many of the pages will not be especially attractive or memorable so try to liven them up a bit with colour as you are reviewing and rewriting. **This is a great memory aid, and memory is the most important thing.**

Activity 2 – Use technology!

Why should everything be written down? Have you thought about "mental" maps, diagrams, cartoons and colour to help you learn? And rather than write down notes, why not record your revision material?

What about having a text message revision session with friends? Keep in touch with them to find out how and what they are revising and share ideas and questions.

Why not make a video diary where you tell the camera what you are doing, what you think you have learned and what you still have to do? No one has to see or hear it, but the process of having to organise your thoughts in a formal way to explain something is a very important learning practice.

Be sure to make use of electronic files. You could begin to summarise your class notes. Your typing might be slow, but it will get faster and the typed notes will be easier to read than the scribbles in your class notes. Try to add different fonts and colours to make your work stand out. You can easily Google relevant pictures, cartoons and diagrams which you can copy and paste to make your work more attractive and **MEMORABLE**.

Activity 3 – This is it. Do this and you will know lots!

Step 1

In this task you must be very honest with yourself! Find the SQA syllabus for your subject (www.sqa.org.uk). Look at how it is broken down into main topics called MANDATORY knowledge. That means stuff you MUST know.

Step 2

BEFORE you do ANY revision on this topic, write a list of everything that you already know about the subject. It might be quite a long list but you only need to write it once. It shows you all the information that is already in your long-term memory so you know what parts you do not need to revise!

Step 3

Pick a chapter or section from your book or revision notes. Choose a fairly large section or a whole chapter to get the most out of this activity.

With a buddy, use Skype, Facetime, Twitter or any other communication you have, to play the game "If this is the answer, what is the question?". For example, if you are revising Geography and the answer you provide is "meander", your buddy would have to make up a question like "What is the word that describes a feature of a river where it flows slowly and bends often from side to side?".

Make up 10 "answers" based on the content of the chapter or section you are using. Give this to your buddy to solve while you solve theirs.

Step 4

Construct a wordsearch of at least 10 × 10 squares. You can make it as big as you like but keep it realistic. Work together with a group of friends. Many apps allow you to make wordsearch puzzles online. The words and phrases can go in any direction and phrases can be split. Your puzzle must only contain facts linked to the topic you are revising. Your task is to find 10 bits of information to hide in your puzzle, but you must not repeat information that you used in Step 3. DO NOT show where the words are. Fill up empty squares with random letters. Remember to keep a note of where your answers are hidden but do not show your friends. When you have a complete puzzle, exchange it with a friend to solve each other's puzzle.

Step 5

Now make up 10 questions (not "answers" this time) based on the same chapter used in the previous two tasks. Again, you must find NEW information that you have not yet used. Now it's getting hard to find that new information! Again, give your questions to a friend to answer.

Step 6

As you have been doing the puzzles, your brain has been actively searching for new information. Now write a NEW LIST that contains only the new information you have discovered when doing the puzzles. Your new list is the one to look at repeatedly for short bursts over the next few days. Try to remember more and more of it without looking at it. After a few days, you should be able to add words from your second list to your first list as you increase the information in your long-term memory.

FINALLY! Be inspired...

Make a list of different revision ideas and beside each one write **THINGS I HAVE** tried, **THINGS I WILL** try and **THINGS I MIGHT** try. Don't be scared of trying something new.

And remember – "FAIL TO PREPARE AND PREPARE TO FAIL!"

National 5 Biology

The Course

The National 5 Biology Course consists of three National Units. These are *Cell Biology*, *Multicellular Organisms* and *Life on Earth*. In each of the Units you will be assessed on your ability to demonstrate and apply knowledge of Biology, and to demonstrate and apply skills of scientific inquiry. Candidates must also complete an Assignment in which they research a topic in biology and write it up as a report. They also take a Course Examination.

How the Course is graded

To achieve a Course award for National 5 Biology you must pass all three National **Unit Assessments**, which will be assessed by your school or college on a pass or fail basis. The grade you get depends on the following two Course assessments, which are set and graded by SQA.

1. **An Assignment** that requires you to write a 500 – 800 word report. The Assignment is 20% of your grade and is marked out of 20 marks, most of which are allocated for skills of scientific inquiry.
2. A written **Course Examination**, which is worth the remaining 80% of the grade. The Examination is marked out of 80 marks, most of which are for the demonstration and application of knowledge, although there are also marks available for skills of scientific inquiry. This book should help you practise the Examination part!

To pass National 5 Biology with a C grade you will need about 50% of the 100 marks available for the Assignment and the Course Examination combined. For a B, you will need 60%, and for an A, 70%.

The Course Examination

The Course Examination is a single question paper split into two sections. The first section is an objective test with 20 multiple choice items for 20 marks. The second section is a mixture of restricted and extended response questions worth between 1 and 3 marks each for a total of 60 marks. Some questions will contain options and there will usually be a question that asks you to suggest changes to experimental methods. Altogether there are 80 marks, and you will have 2 hours to complete the paper. Most of the marks are for knowledge and its application, with the remainder of questions designed to test skills of scientific inquiry.

The majority of the marks will be straightforward – these are the marks that will help you get a grade C. Some questions will be more demanding – these are the questions you need to get right to get a grade A.

General hints and tips

You should have a copy of the Course Assessment Specification for National 5 Biology – if you haven't got one, download it from the SQA website. This document tells you what you may be tested on in your examination. It is worth spending some time studying this document.

This book contains three past National 5 Biology examination papers. Notice how similar they all are in the way in which they are laid out and the types of question they ask – your own Course Examination will be very similar as well so working through the papers in this book will be good preparation.

If you are trying a whole examination paper from this book, give yourself a maximum of two hours to complete it. The questions in each paper are laid out in Unit order. Make sure that you spend time using the answer section to mark your own work – it is especially useful if you can get someone to help you with this. You could even grade your work on an A–D basis.

The following hints and tips are related to examination techniques as well as avoiding common mistakes.

Remember that if you hit problems with a question, you should ask your teacher for help.

Section 1

20 multiple-choice items — 20 marks

- Answer on a grid.
- Do not spend more than **30 minutes** on this section.
- Some individual questions might take longer to answer than others – this is quite normal and make sure you use scrap paper if a calculation or any working is needed.
- Some questions can be answered instantly – again, this is normal.
- **Do not leave blanks** – complete the grid for each question as you work through.
- Try to answer each question in your head **without** looking at the options. If your answer is there – you are home and dry!

- If you are not certain, choose the answer that seemed most attractive on **first** reading the answer options.
- If you are guessing, try to eliminate options before making your guess. If you can eliminate 3 – you are left with the correct answer even if you do not recognise it!

Section 2

Restricted and extended response 60 marks

- Spend about **90 minutes** on this section.
- Answer on the question paper. Try to write neatly and keep your answers on the support lines if possible – the lines are designed to take the full answer!
- A clue to answer length is the mark allocation – most questions are restricted to 1 mark and the answer can be quite short. If there are 2 or 3 marks available, your answer will need to be extended and may well have two, three or even four parts.
- The questions are usually laid out in Unit sequence but remember some questions are **designed** to cover more than one Unit.
- The grade C-type questions usually start with "**State**", "**Identify**", "**Give**" or "**Name**" and often need only a word or two in response. They will usually be worth one mark each.
- Questions that begin with "**Explain**" and "**Describe**" are usually grade A types and are likely to have more than one part to the full answer. You will usually have to write a sentence or two and there may be two or even three marks available.
- Make sure you read questions through twice before trying to answer – there is often very important information within the question.
- Using abbreviations like DNA and ATP is fine, and the bases of DNA can be given as A, T, G and C.
- Don't worry that a few questions are in unfamiliar contexts – that's the idea! Just keep calm and read the questions carefully.
- If a question contains a choice, be sure to spend a minute or two making the best choice for you.
- In experimental questions, you must be aware of what variables are, why controls are needed and how reliability might be improved. It is worth spending time on these ideas – they are essential and will come up year after year.
- Some candidates like to use a highlighter pen to help them focus on the essential points of longer questions – this is a great technique.
- Remember that a **conclusion** can be seen from data, whereas an **explanation** will usually require you to supply some background knowledge as well.
- Remember to "**use values from the graph**" when describing graphical information in words if you are asked to do so.
- Plot graphs carefully and join the plot points using a ruler. Include zeros on your scale where appropriate and use the data table headings for the axes labels.
- Look out for graphs with two Y axes – these need extra special concentration and anyone can make a mistake!
- If you are given space for a calculation you will very likely need to use it! A calculator is essential.
- The main types of calculation tend to be **ratios**, **averages** and **percentages** – make sure you can do these common calculations.
- Answers to calculations will not usually have more than two decimal places.
- Do not leave blanks. Always have a go, using the language in the question if you can.

Good luck!

Remember that the rewards for passing National 5 Biology are well worth it! Your pass will help you get the future you want for yourself. In the exam, be confident in your own ability. If you're not sure how to answer a question, trust your instincts and just give it a go anyway. Keep calm and don't panic! GOOD LUCK!

NATIONAL 5

2014

X707/75/02

Biology
Section 1—Questions

FRIDAY, 16 MAY

9:00 AM – 11:00 AM

Instructions for the completion of Section 1 are given on Page two of your question and answer booklet X707/75/01.

Record your answers on the answer grid on Page three of your question and answer booklet.

Before leaving the examination room you must give your question and answer booklet to the Invigilator; if you do not, you may lose all the marks for this paper.

SECTION 1

1. Which structural feature is found in a plant cell and not in an animal cell?

   A Nucleus

   B Cell wall

   C Cell membrane

   D Cytoplasm

2. Which line in the table below identifies the direction of diffusion of the three substances during muscle contraction?

	Substance		
	Glucose	*Oxygen*	*Carbon dioxide*
A	out	out	in
B	in	out	in
C	out	in	out
D	in	in	out

3. The diagram below represents a genetically engineered bacterial cell.

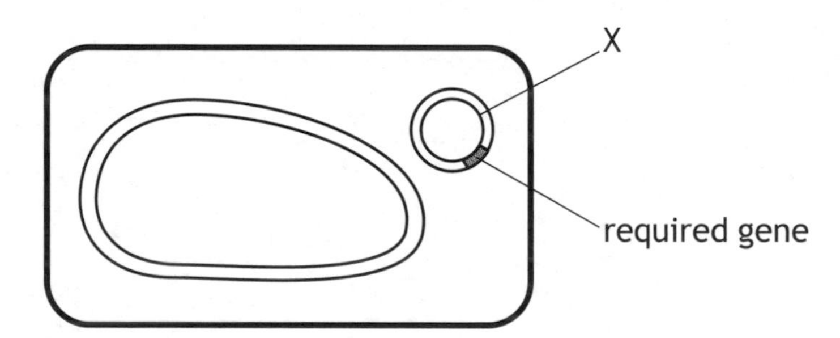

The structure labelled X is a

   A chromosome

   B plasmid

   C ribosome

   D nucleus.

4. The light energy for photosynthesis is captured by

A water

B hydrogen

C chlorophyll

D oxygen.

5. The diagram below represents the human brain.

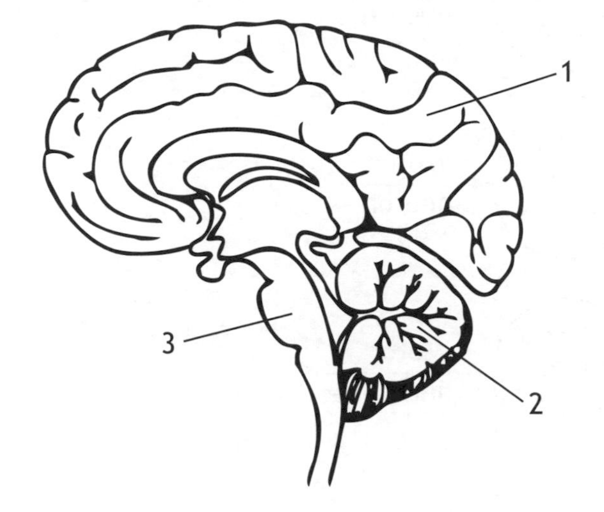

Which line in the table below identifies structures 1, 2 and 3 of the human brain?

	Structure 1	*Structure 2*	*Structure 3*
A	medulla	cerebrum	cerebellum
B	cerebrum	medulla	cerebellum
C	cerebellum	cerebrum	medulla
D	cerebrum	cerebellum	medulla

6. Proteins have different functions. Which of the following statements identifies a **protein** and its function?

A Hormones carry oxygen around the body.

B Enzymes carry chemical messages around the body.

C Antibodies defend the body against disease.

D Cellulose provides strength and structure to a plant cell wall.

[Turn over

7. Which of the diagrams below identifies neurons and the direction of travel of nerve impulses?

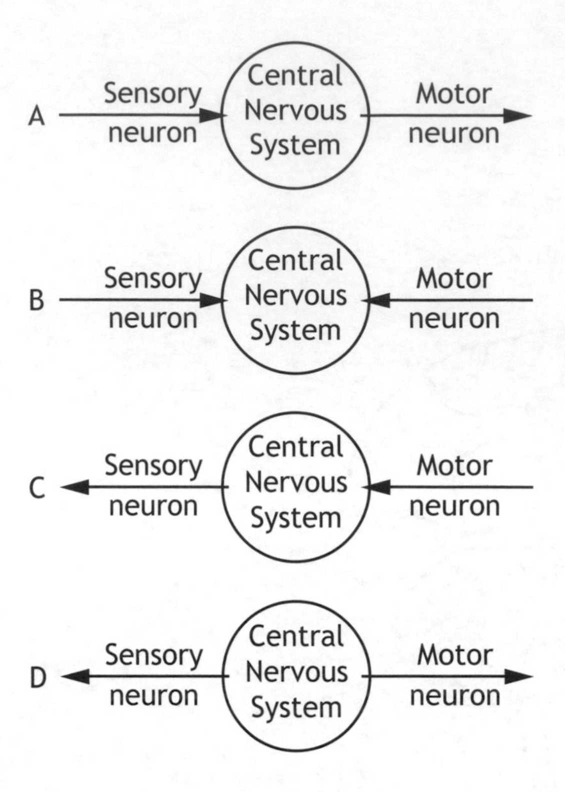

8. Which of the following pairs of human cells have the same number of chromosomes?

A Liver cell and sperm cell

B Kidney cell and sperm cell

C Kidney cell and liver cell

D Liver cell and egg cell

9. The table below shows the results of an investigation into the effect of temperature on egg laying in adult red spider mites.

Feature	*Temperature* (°C)		
	20 °C	25 °C	30 °C
Average length of egg laying period (days)	24	18	12
Average number of eggs laid per female during egg laying period	72	72	72

As the temperature increases, the average number of eggs laid per female per day

A increases

B decreases

C stays the same

D halves.

10. The following diagrams show a cell at four different stages of mitosis.

1

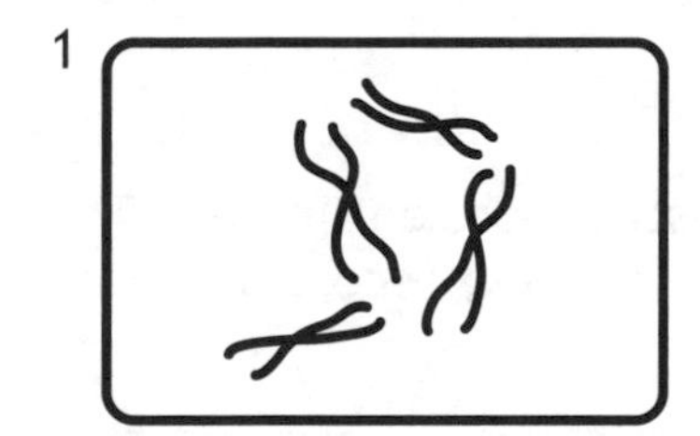

2

3

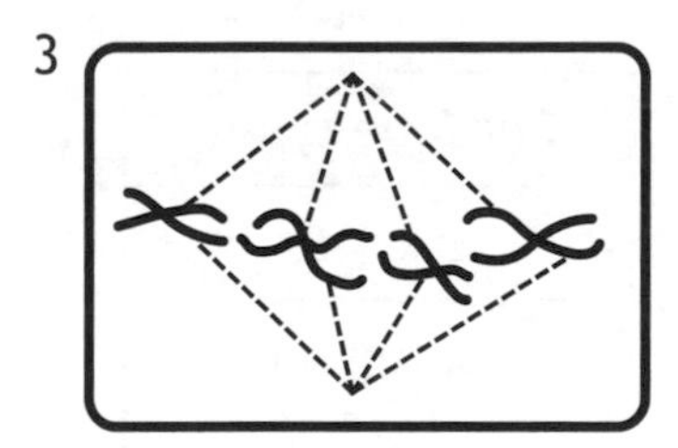

4

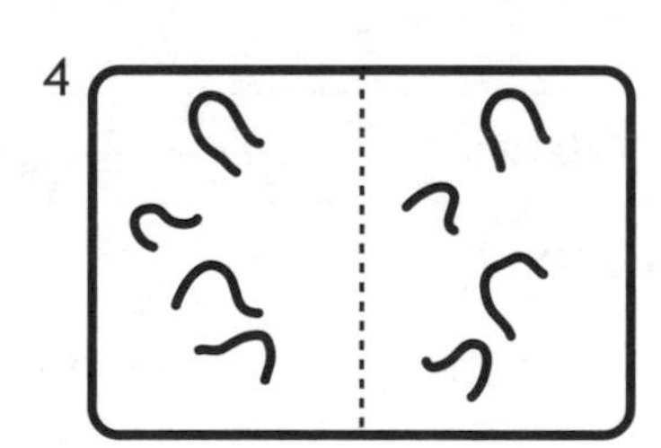

The correct order of the stages of mitosis is

A 1,3,2,4

B 2,3,4,1

C 3,2,1,4

D 4,1,2,3.

11. Which of the following diagrams represents the process of fertilisation in plants?

A
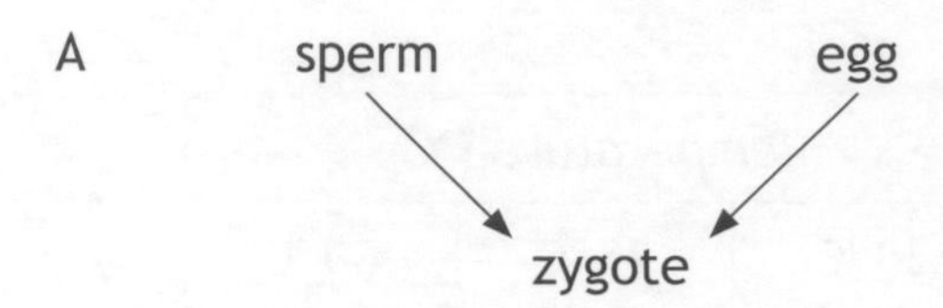

B
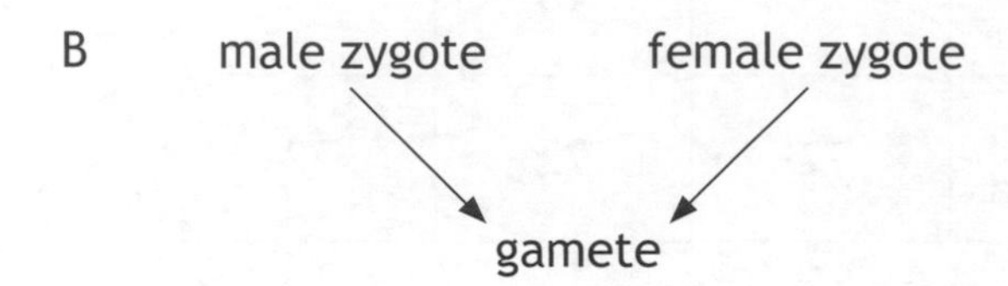

C
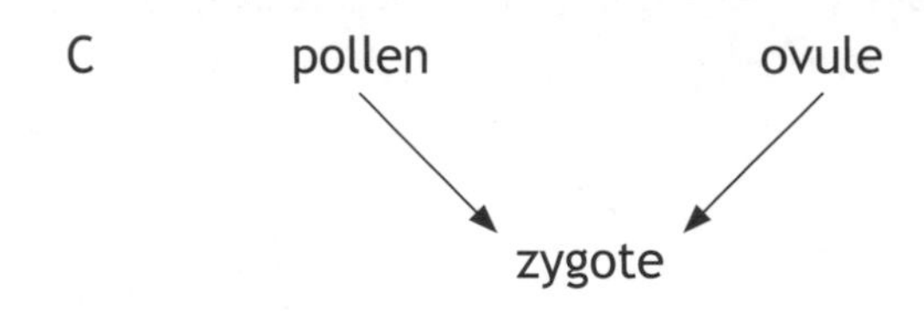

D
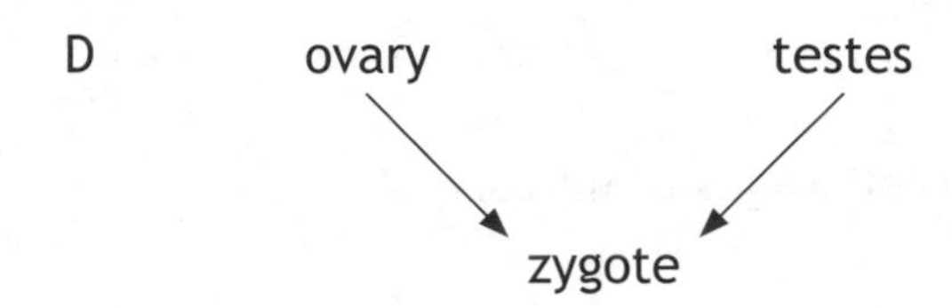

12. Variation in a characteristic can either be discrete or continuous. The range of heights and weights for a group of students were measured and recorded. Ear lobe types were also examined and categorised into groups.

Which line in the table below identifies the type of variation shown by each of these human characteristics?

	Height	*Weight*	*Ear lobe types*
A	continuous	continuous	discrete
B	discrete	continuous	continuous
C	discrete	discrete	continuous
D	continuous	discrete	discrete

13. The diagram below shows part of the human respiratory system.

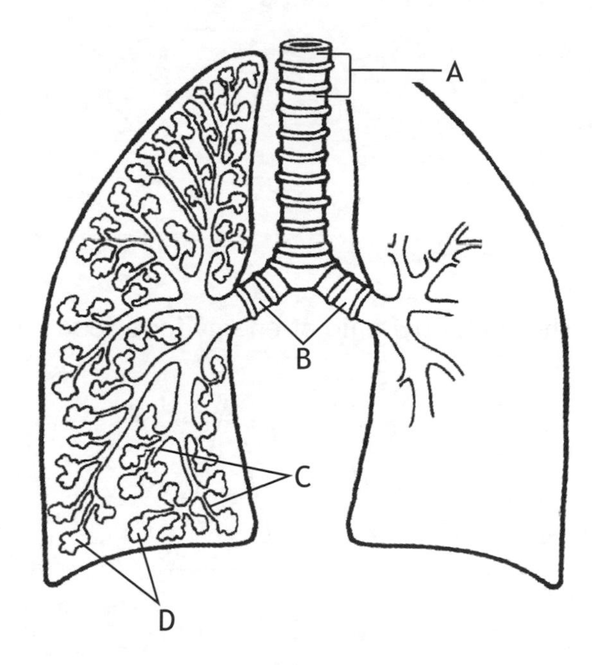

Which letter identifies the alveoli?

14. Which line in the table below identifies abiotic and biotic factors?

	Abiotic factor	*Biotic factor*
A	light intensity	pH
B	temperature	predation
C	grazing	light intensity
D	predation	grazing

[Turn over

15. A rabbit feeds on grass, is eaten by foxes and is a habitat for fleas.

The statement above describes the rabbit's

A ecosystem

B community

C niche

D prey.

16. The diagram below shows the pyramid of energy for a food chain.

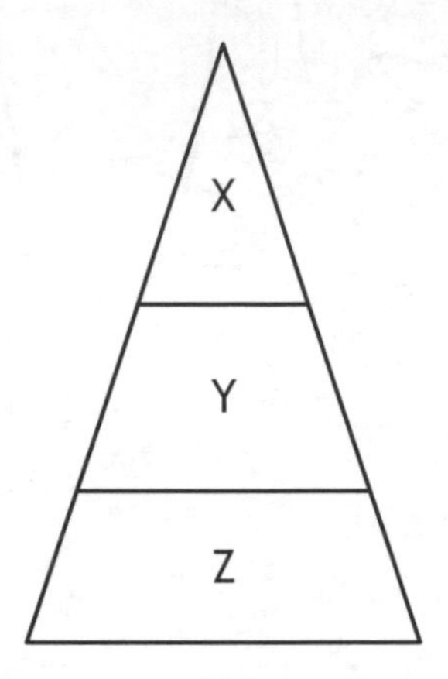

There is a lot less energy at level X in the pyramid because

A there are fewer organisms at this level

B energy is stored at each level

C energy is lost at each level

D the organisms are bigger at this level.

17. In which of the following would competition **not** occur?

A Rabbits grazing in a field

B Owls and foxes hunting for mice

C Daisies and dandelions growing in a lawn

D Algae and fish in a loch

18. The following diagrams represent part of the nitrogen cycle. Which diagram shows the correct sequence of events in the nitrogen cycle?

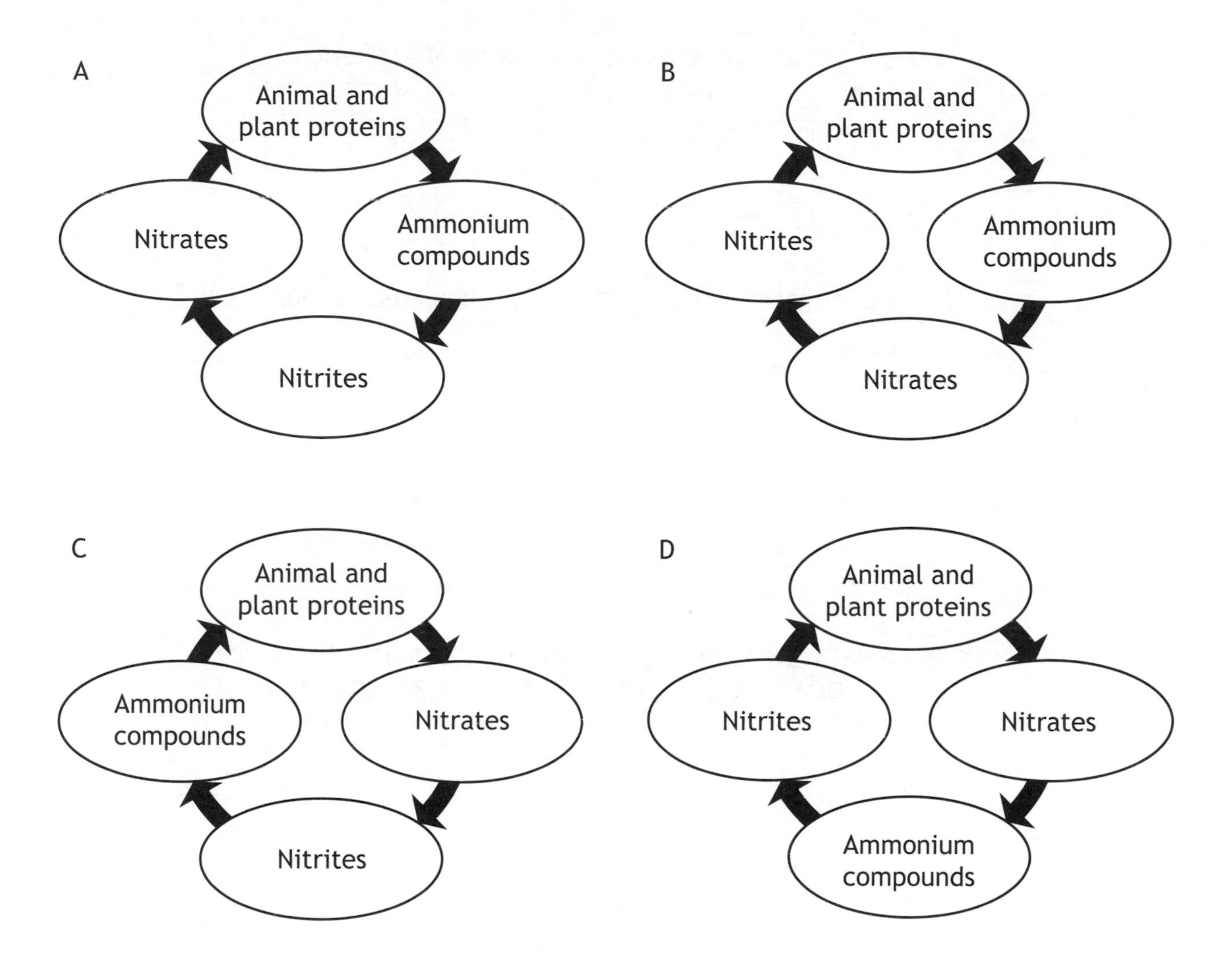

19. Students used a quadrat to estimate the number of buttercups in a field.

They threw the quadrat randomly three times in the area.

In order to improve the reliability of their results they could have

A asked another group of students to check that they had counted correctly

B thrown the quadrat ten times instead of three

C only thrown the quadrat when conditions were at an optimum

D used a smaller quadrat for each of their samples.

[Turn over for Question 20 on *Page ten*

20. The table below compares the rate of extinction of mammal species over two different time periods.

Time period (years)	*Rate of extinction per 100 years*
1500 – 1900	4·5
1900 – 2000	90

The ratio of extinction rates between 1900 – 2000 compared to 1500 – 1900 is

A 1:20

B 1:2

C 2:1

D 20:1.

[END OF SECTION 1. NOW ATTEMPT THE QUESTIONS IN SECTION 2 OF YOUR QUESTION AND ANSWER BOOKLET]

FOR OFFICIAL USE

N5

National Qualifications 2014

Mark

X707/75/01

Biology
Section 1—Answer Grid and Section 2

FRIDAY, 16 MAY

9:00 AM – 11:00 AM

Fill in these boxes and read what is printed below.

Full name of centre

Town

Forename(s)

Surname

Number of seat

Date of birth

Day Month Year

D D M M Y Y

Scottish candidate number

Total marks — 80

SECTION 1 — 20 marks

Attempt ALL questions in this section.

Instructions for the completion of Section 1 are given on Page two.

SECTION 2 — 60 marks

Attempt ALL questions in this section.

Write your answers clearly in the spaces provided in this booklet. Additional space for answers and rough work is provided at the end of this booklet. If you use this space you must clearly identify the question number you are attempting. Any rough work must be written in this booklet. You should score through your rough work when you have written your final copy.

Use **blue** or **black** ink.

Before leaving the examination room you must give this booklet to the Invigilator; if you do not, you may lose all the marks for this paper.

SECTION 1— 20 marks

The questions for Section 1 are contained in the question paper X707/75/02.
Read these and record your answers on the answer grid on Page three opposite.
Do NOT use gel pens.

1. The answer to each question is **either** A, B, C or D. Decide what your answer is, then fill in the appropriate bubble (see sample question below).

2. There is **only one correct** answer to each question.

3. Any rough working should be done on the additional space for answers and rough work at the end of this booklet.

Sample Question

The thigh bone is called the

A humerus

B femur

C tibia

D fibula.

The correct answer is **B**—femur. The answer **B** bubble has been clearly filled in (see below).

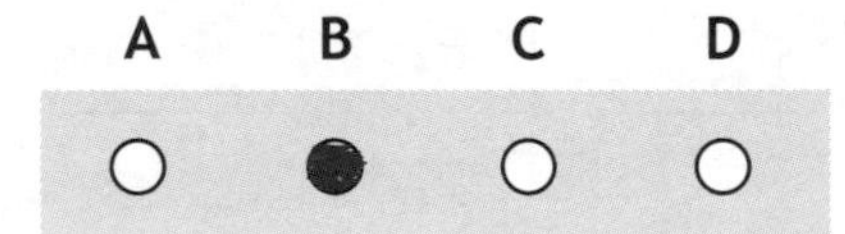

Changing an answer

If you decide to change your answer, cancel your first answer by putting a cross through it (see below) and fill in the answer you want. The answer below has been changed to **D**.

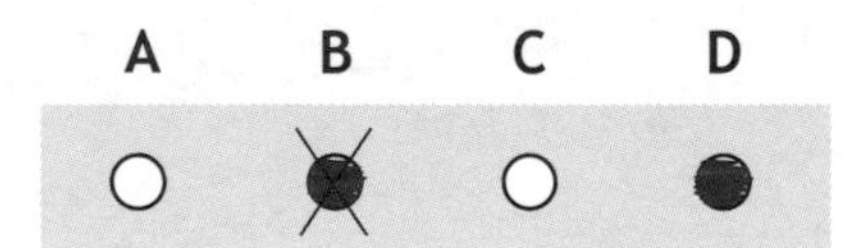

If you then decide to change back to an answer you have already scored out, put a tick (✓) to the **right** of the answer you want, as shown below:

A B C D
○ ⊗✓ ○ ⊗

or

A B C D
○ ⊗✓ ○ ○

SECTION 1 — Answer Grid

	A	B	C	D
1	○	○	○	○
2	○	○	○	○
3	○	○	○	○
4	○	○	○	○
5	○	○	○	○
6	○	○	○	○
7	○	○	○	○
8	○	○	○	○
9	○	○	○	○
10	○	○	○	○
11	○	○	○	○
12	○	○	○	○
13	○	○	○	○
14	○	○	○	○
15	○	○	○	○
16	○	○	○	○
17	○	○	○	○
18	○	○	○	○
19	○	○	○	○
20	○	○	○	○

[BLANK PAGE]

DO NOT WRITE ON THIS PAGE

[Turn over for Question 1 on *Page six*

DO NOT WRITE ON THIS PAGE

MARKS | DO NOT WRITE IN THIS MARGIN

SECTION 2 — 60 marks

Attempt ALL questions

1. A group of students carried out an investigation into the variety of cell types.

The types of cell they examined are shown in the box below.

Animal	Plant	Bacterial	Fungal

(a) (i) Identify the type(s) of cell which have a cell wall. **1**

____

(ii) Identify the type(s) of cell which have a plasmid. **1**

____

(iii) Some organelles are found in all cells.

Choose one of the following organelles and tick (✓) the appropriate box.

Describe the function of the chosen organelle. **1**

Ribosome ☐ Mitochondria ☐

Function ____

____

MARKS DO NOT WRITE IN THIS MARGIN

1. (continued)

(b) The students then measured a number of cells and calculated the average cell sizes. The results are shown in the table below.

Type of cell	*Average size of cell (μm)*
Animal	24
Plant	48
Bacterial	3
Fungal	7

On the graph paper below, complete the vertical axis and draw a bar chart to show the average size of the cells shown in the table. **2**

(Additional graph paper, if required, can be found on *Page twenty-six*)

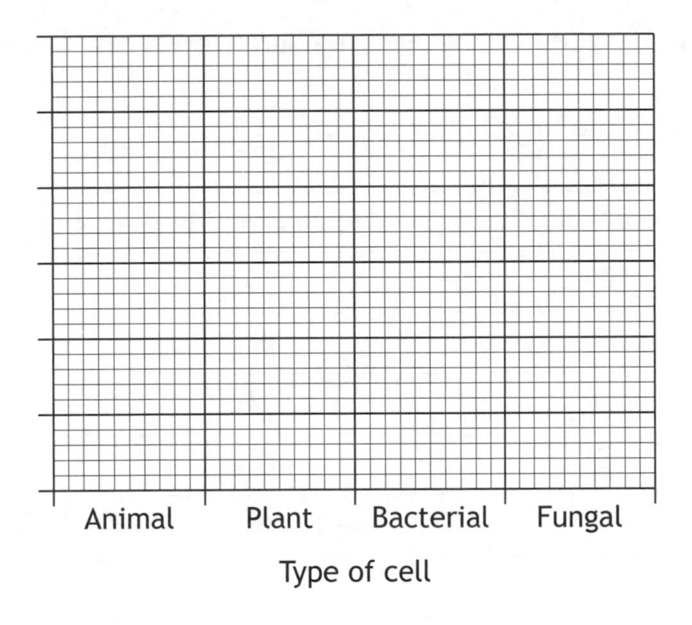

Total marks 5

[Turn over

MARKS | DO NOT WRITE IN THIS MARGIN

2. The apparatus shown below was used to investigate the movement of water into and out of a model cell. The model cell had a selectively permeable membrane.

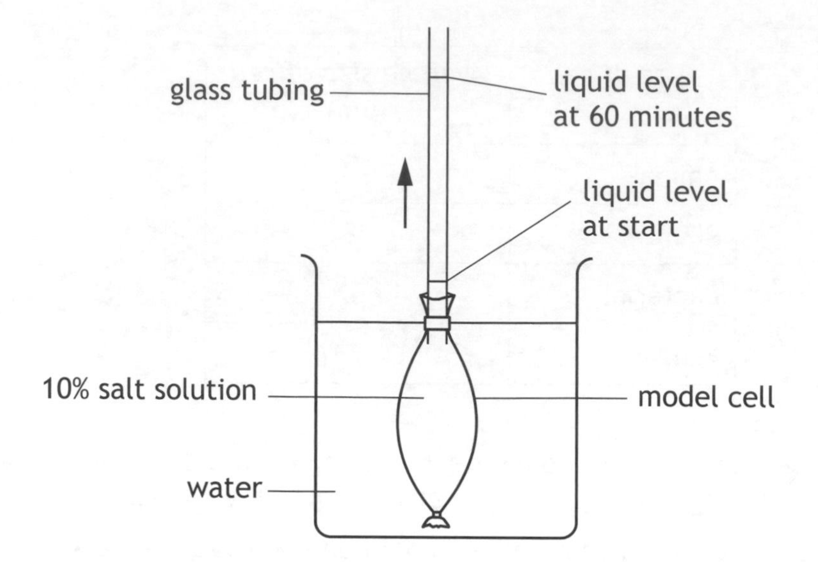

The liquid level in the glass tubing was measured every 10 minutes for 60 minutes.

The results are shown in the table below.

Time (minutes)	*Liquid level* (mm)
0	10
10	22
20	32
30	40
40	48
50	56
60	64

(a) Name the process which caused the liquid level to rise. 1

____

MARKS DO NOT WRITE IN THIS MARGIN

2. (continued)

(b) Explain how this process caused the liquid level to rise. 2

____

____

____

(c) Calculate the average rate of movement of liquid in the glass tubing. 1

Space for calculation

_______________ mm per minute

(d) When the investigation was repeated, the average rate of movement of liquid was slower.

Suggest **one** difference in the way that the investigation was set up that could have caused this change in results. 1

____

____

Total marks 5

[Turn over

MARKS | DO NOT WRITE IN THIS MARGIN

3. (a) Hydrogen peroxide can damage cells and lead to cell death. Catalase is an enzyme which breaks down hydrogen peroxide into oxygen and water.

Scientists in New Zealand investigated the link between the level of catalase in sheep livers and the fat in their meat. The hypothesis was that the higher the level of liver catalase, the greater the fat content of the meat.

In the investigation, they examined 9 sheep with a high percentage of fat and 15 sheep with a low percentage of fat. The sheep with the high percentage of fat had an average catalase level of 4800 K/g and those with the lower percentage of fat had an average catalase level of 3600 K/g.

The scientists concluded that their hypothesis was correct.

(i) Name the substrate of catalase. **1**

(ii) Identify an aspect in the planning of the investigation that would suggest that the hypothesis might not be proven correct. **1**

(iii) A further investigation proved that the hypothesis was correct.

Describe how this investigation could help farmers to select only sheep with a low percentage of fat, to provide meat for consumers following a low fat diet. **1**

(b) The optimum temperature for the activity of catalase is 37 °C.

Predict what would happen to the activity of catalase if the temperature was lowered to 34 °C. **1**

Total marks 4

MARKS DO NOT WRITE IN THIS MARGIN

4. The following diagram shows a cross-section of some villi in the small intestine.

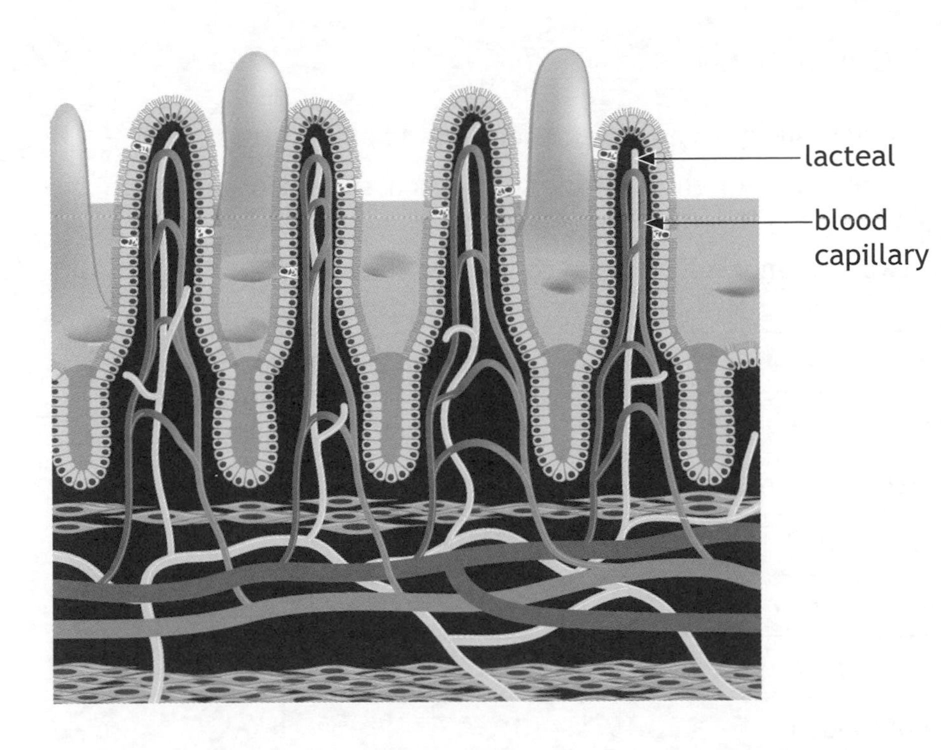

Explain why the **structure and number** of villi make absorption an efficient process in the small intestine. 3

____

____

____

____

____

____

[Turn over

MARKS DO NOT WRITE IN THIS MARGIN

5. Photosynthesis is a two stage process.

Stage 1 — Light reactions

Stage 2 — Carbon fixation

(a) The table below shows some statements about photosynthesis.

Complete the table to show which stage each statement refers to by placing a tick (✓) in the Stage 1 or Stage 2 box.

The first two statements have been completed for you. **2**

Statement	*Stage 1*	*Stage 2*
Carbon dioxide required		✓
Light energy required	✓	
Water required		
Sugar produced		
ATP + Hydrogen required		
Oxygen produced		

(b) Explain why high temperatures (above 50°C) would prevent the photosynthesis reactions from taking place. **2**

____

____

____

MARKS | DO NOT WRITE IN THIS MARGIN

5. **(continued)**

(c) The graph below shows how the rate of photosynthesis is affected by the concentration of carbon dioxide.

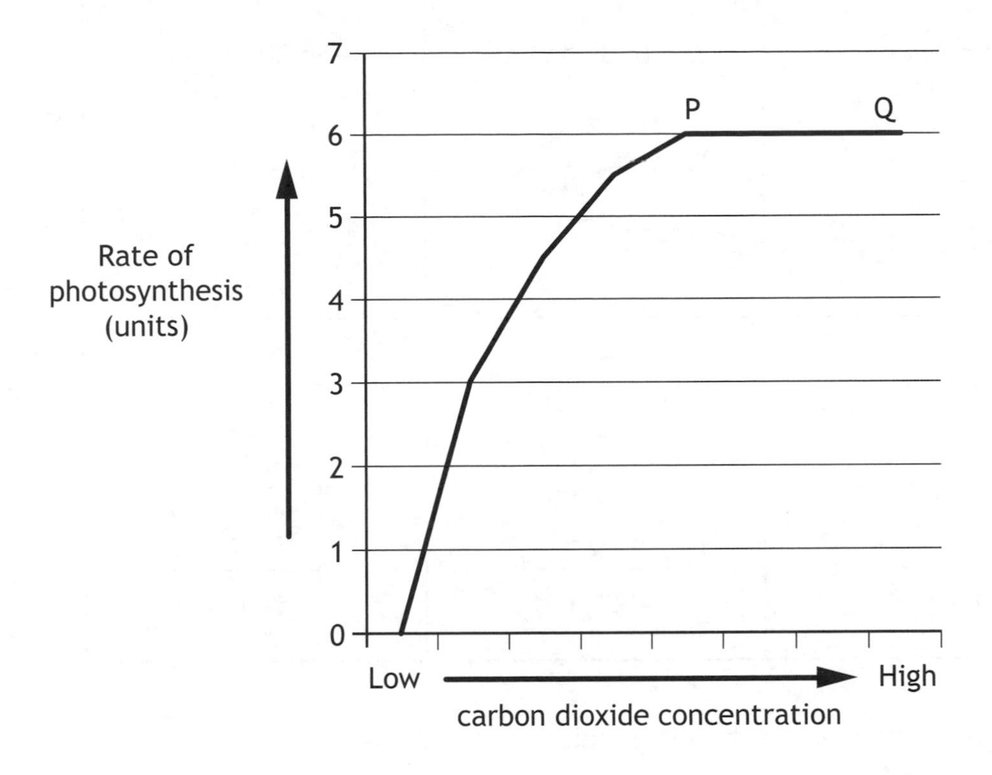

State two environmental factors which could limit the rate of photosynthesis between points P and Q. **1**

1 ____________________

2 ____________________

Total marks 5

[Turn over

MARKS | DO NOT WRITE IN THIS MARGIN

6. The diagrams below show examples of some types of specialised cells from the human body.

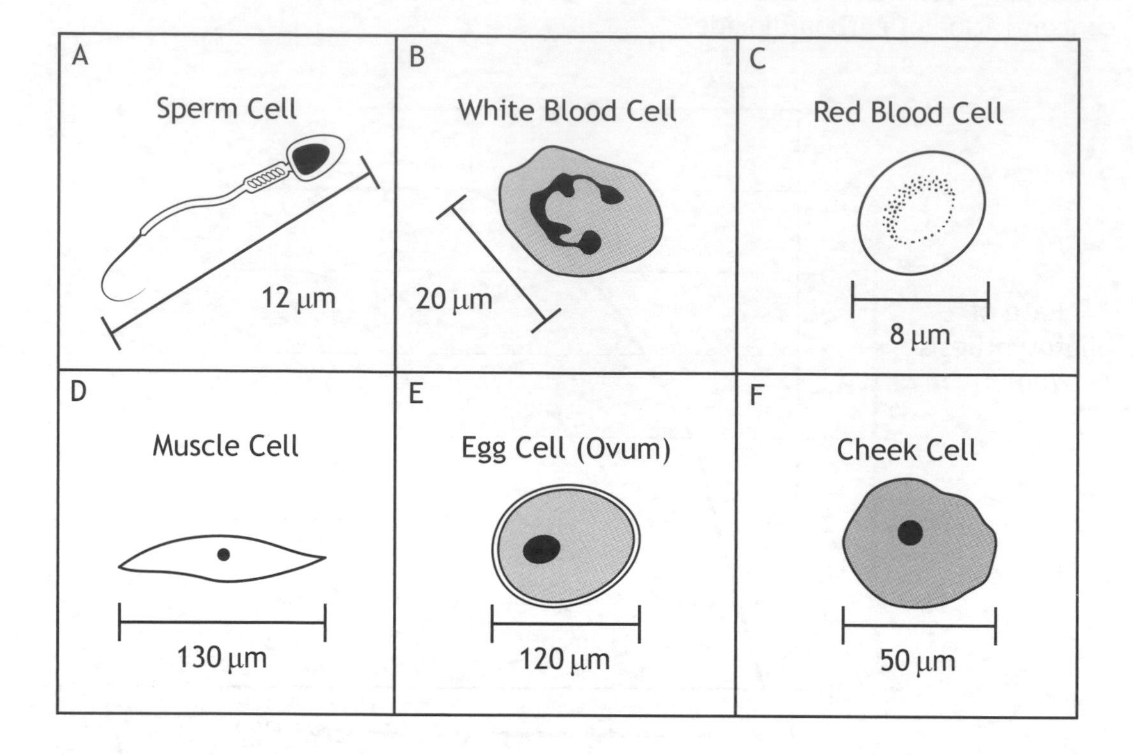

The cells are not drawn to the same scale.
(μm = micrometre)

(a) Put letters in the boxes below to arrange the cells in order of size. **1**

increasing size →

(b) Choose one of the following cell types by (circling) it.

sperm cell   egg cell   red blood cell

Describe the function of the chosen cell and explain how its specialisation allows it to carry out that function. **2**

Function ______________________________

Explanation ______________________________

____

MARKS | DO NOT WRITE IN THIS MARGIN

6. **(continued)**

(c) The diagram below shows some stages in the development of blood cells and nerve cells.

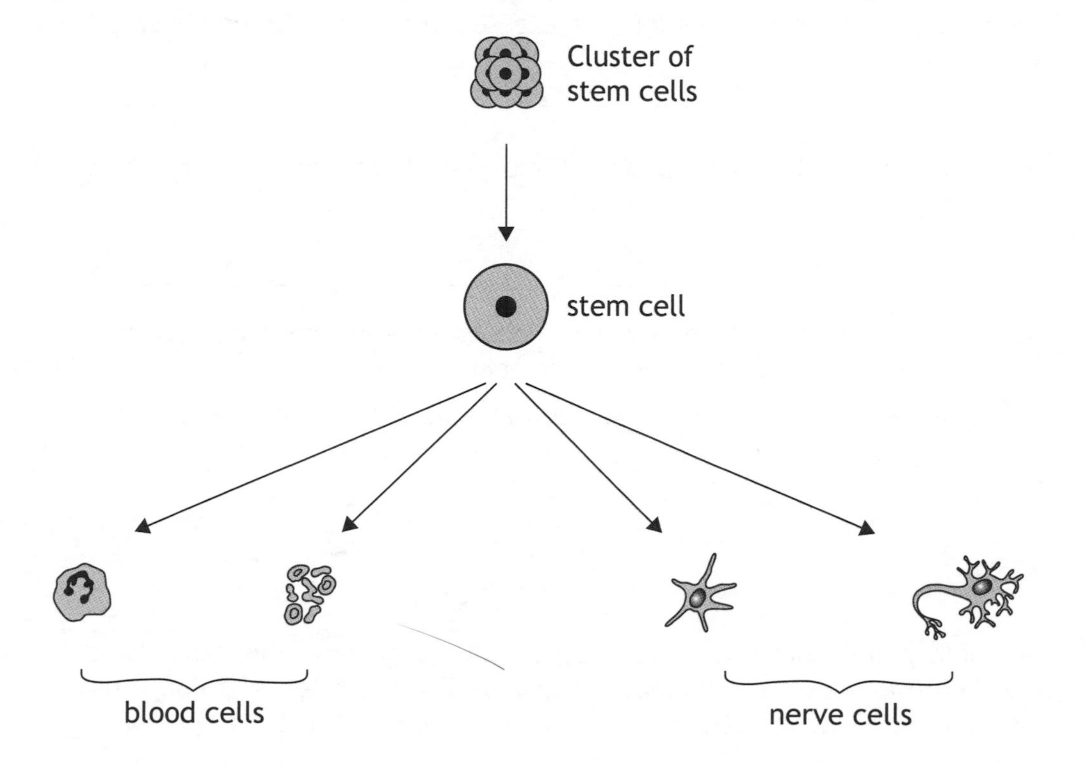

Describe the feature of stem cells which gives them the potential to develop into many different types of cells, such as blood and nerve cells. **1**

____

(d) Which of the following statements refer to processes involving stem cells?

Tick (✓) the correct box(es). **1**

Growth of new skin	☐
Transmission of nerve impulses	☐
Muscle contraction	☐
Repair of broken bones	☐
Production of insulin	☐

Total marks 5

MARKS DO NOT WRITE IN THIS MARGIN

7. Muscle tissue can be dark or light in colour.

Dark tissue cells use oxygen to release energy.

Light tissue cells do not use oxygen to release energy.

(a) Name the process by which energy is released in the dark tissue cells. **1**

____

(b) (i) Name the substance which muscle cells break down to produce pyruvate. **1**

____

(ii) When pyruvate is being formed, enough energy is released to form two molecules of a high energy compound.

Complete the word equation below to show how this compound is generated. **1**

__________ + __________ ⟶ __________

(c) The table below shows the average percentage of dark and light tissue cells. These cells were found in the muscles of athletes training for different events at the 2014 Commonwealth games in Scotland.

Type of Athlete	*Average percentage of dark tissue cells* (%)	*Average percentage of light tissue cells* (%)
cyclist	60	40
swimmer	75	25
shot putter	40	60
marathon runner	82	18
sprinter	38	62

(i) Using information in the table, identify which type of athlete would be likely to produce the most lactic acid in their muscle cells. Justify your answer. **2**

Type of athlete ______________________________

Justification ______________________________

____

MARKS DO NOT WRITE IN THIS MARGIN

7. (continued)

(ii) A sample of muscle tissue from an athlete was examined and found to contain a total of 360 cells.

90 of these cells were light tissue cells.

Identify which type of athlete the sample was taken from. **1**

Space for calculation

Type of athlete ________________________________

Total marks 6

[Turn over

MARKS | DO NOT WRITE IN THIS MARGIN

8. (a) The regulation of glucose in the blood is represented in the diagram below.

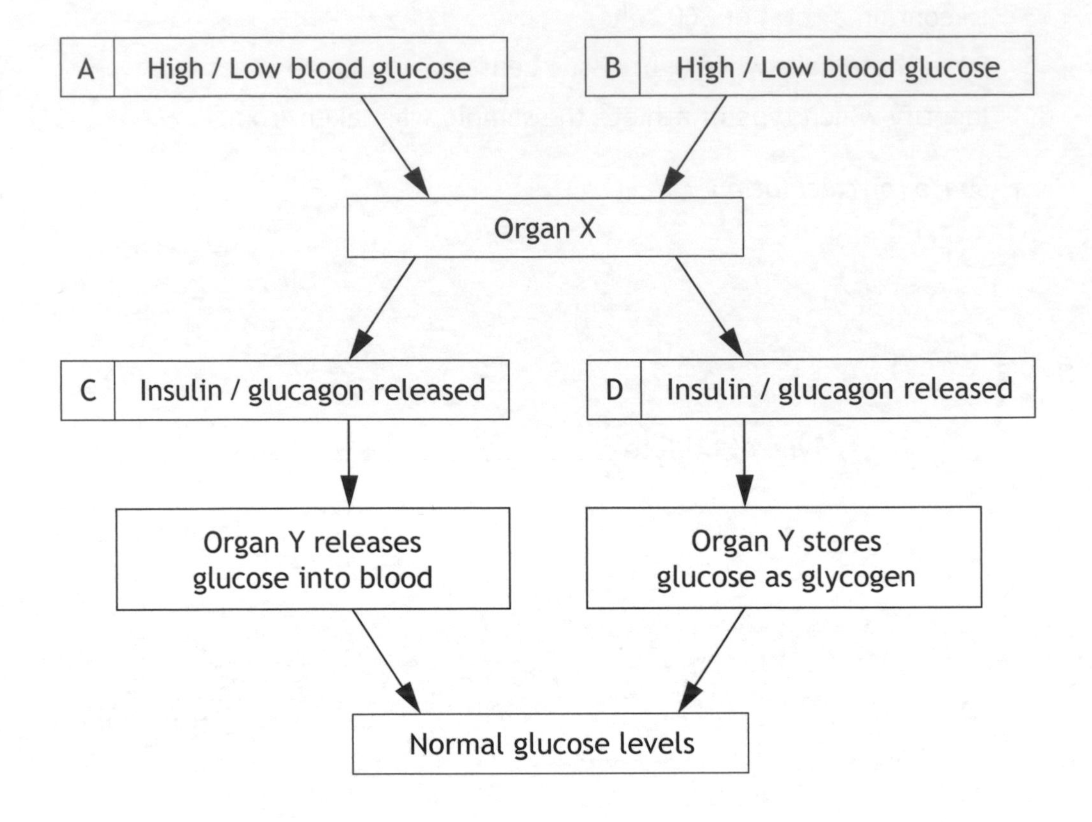

(i) The diagram above has two options in each of the four boxes A, B, C, D.

Circle the correct option in each box. **2**

(ii) Identify organs X and Y. **2**

Organ X ______________________________

Organ Y ______________________________

(b) Insulin and glucagon are hormones.

Describe two features of hormones. **2**

1 ______________________________

2 ______________________________

Total marks 6

MARKS | DO NOT WRITE IN THIS MARGIN

9. Coat colour in Labrador dogs is an inherited characteristic. Black coat (**B**) colour is dominant to chocolate coat colour (**b**).

(a) A homozygous black Labrador was crossed with a Labrador with a chocolate coloured coat.

Complete the diagram below to show the genotypes of each of the parents and the F_1 phenotype. **2**

Parents:	black coat	X	chocolate coat
Genotypes:			
F_1 genotype:		All Bb	
F_1 phenotype:			

(b) (i) Explain what is meant by polygenic inheritance. **1**

____

____

(ii) State the type of variation shown by polygenic inheritance. **1**

____

Total marks 4

[Turn over

MARKS | DO NOT WRITE IN THIS MARGIN

10. (a) Lugworms live on the seashore in dark moist burrows under the sand.

The graph below shows the average number of lugworms at different distances from the seawater at low tide.

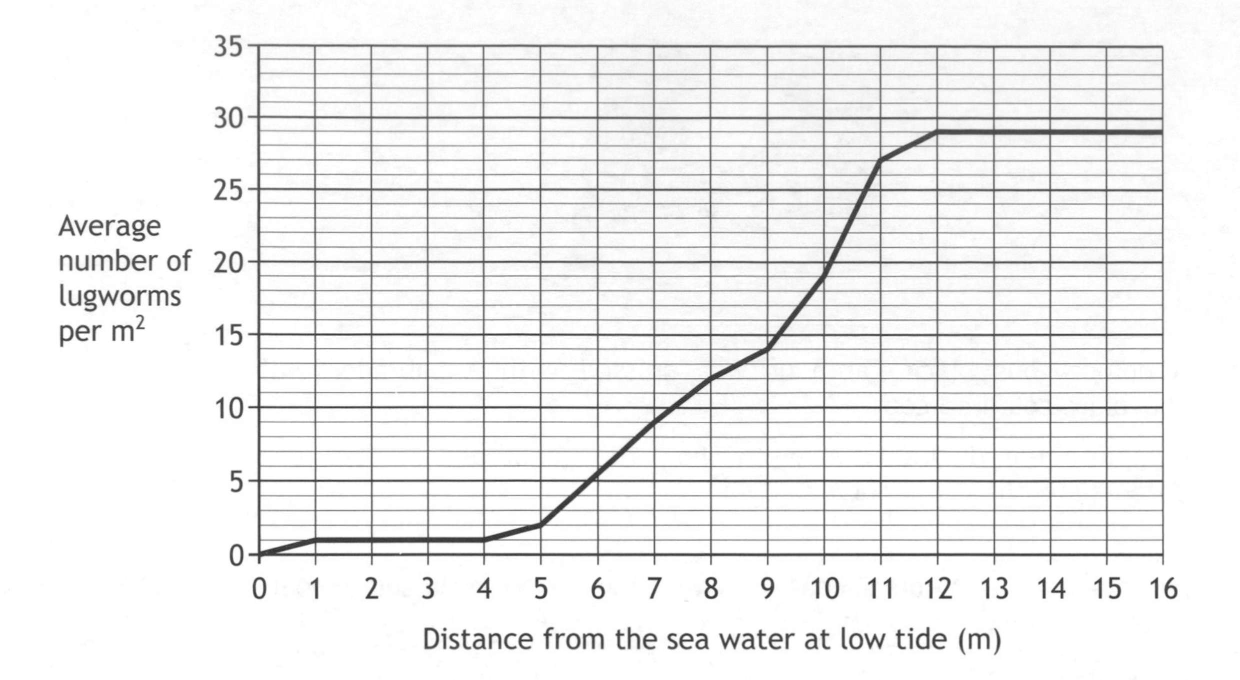

(i) Describe the relationship between the distance from the seawater at low tide and the average number of lugworms per m^2. **2**

____

____

____

____

(ii) Calculate how many times greater the average number of lugworms at 11 metres is compared to 7 metres from the seawater at low tide. **1**

Space for calculation

______________ times greater

MARKS | DO NOT WRITE IN THIS MARGIN

10. (continued)

(b) Dover sole and rex sole are different species of flatfish and are predators of lugworms. Curlews, which are a species of wading bird, also feed on lugworms.

(i) Complete the table below by placing a tick (✓) in the correct box to show the type of competition that would occur between the different predators. **1**

Predator	*Type of Competition*	
	Intraspecific	*Interspecific*
rex sole and curlew		
curlew and curlew		
rex sole and dover sole		

(ii) A curlew gains an average of 165 kilojoules (kJ) of energy daily, by feeding on lugworms.

Select, from the following list, the value of the energy which is used for growth each day by the curlew.

Tick (✓) the correct box. **1**

165 kJ ☐

148·5 kJ ☐

16·5 kJ ☐

0 kJ ☐

Total marks 5

[Turn over

MARKS DO NOT WRITE IN THIS MARGIN

11. During a woodland survey, a group of students measured some abiotic factors. Readings they took included the temperature of the soil and the air.

(a) Name one abiotic factor, other than temperature, which they could have measured in the woodland and describe the method of measuring this factor. **2**

Abiotic factor ______________________________

Method ______________________________

____

____

(b) (i) During the survey, the students sampled the leaf litter in the woodland using pitfall traps.

However, when they checked the pitfall traps four days after setting them up, the students discovered that they were all empty.

Describe an error the students might have made which would explain why there were no invertebrates in the traps. **1**

____

____

____

MARKS DO NOT WRITE IN THIS MARGIN

11. (b) (continued)

(ii) The error was corrected and the students set out the pitfall traps once again. The table below shows the types of invertebrates and numbers found.

Invertebrates	*Number found*
Woodlice	35
Beetles	20
Slugs	0
Spiders	30
Snails	15

Use the information in the table to complete the pie chart below. **2**

(An additional pie chart, if required, can be found on *Page twenty-six*.)

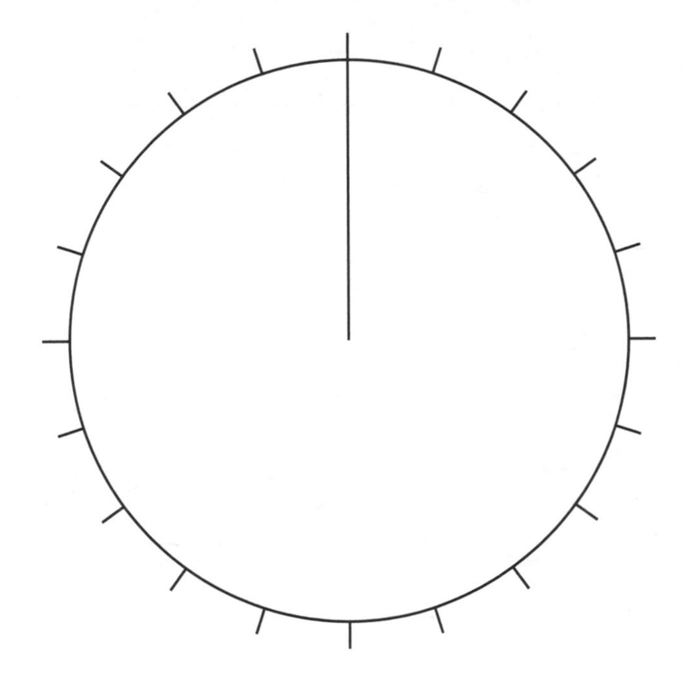

(c) The students saw a large number of butterflies in the woodland.

Give a reason why no butterflies were collected with the invertebrates. **1**

____

____

Total marks 6

MARKS | DO NOT WRITE IN THIS MARGIN

12. The following diagram shows the stages in the formation of a new species.

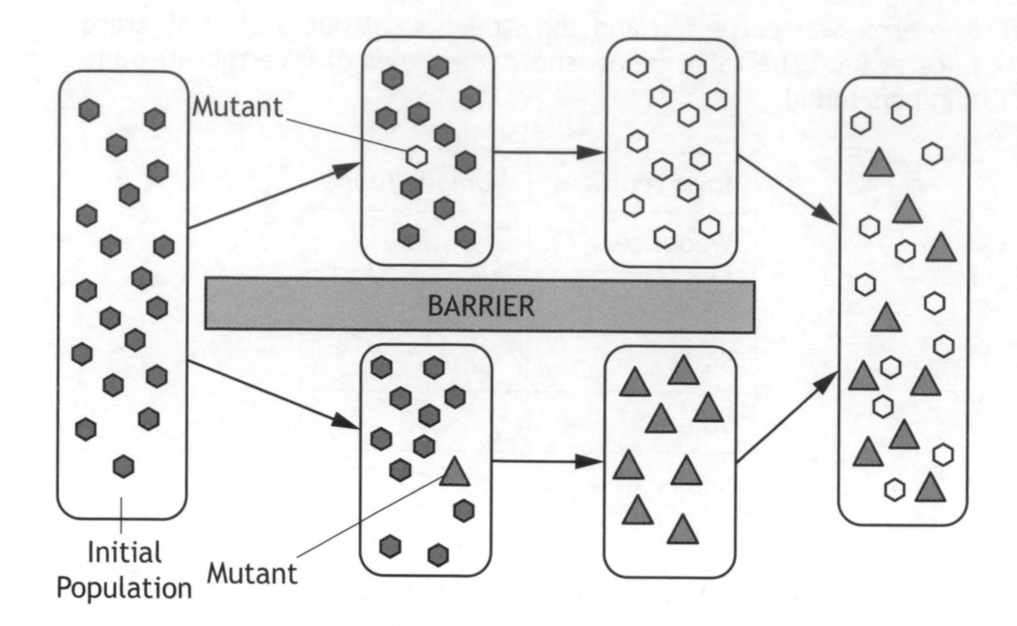

(a) Using the information in the diagram, describe how new species are formed. **4**

MARKS | DO NOT WRITE IN THIS MARGIN

12. (continued)

(b) Choose either mutation or species and tick (✓) the appropriate box.
Give a definition of the chosen term. **1**

Mutation ☐ Species ☐

Definition ______________________________

____

____

(c) In any population, variation exists. Explain why variation is important for the survival of a population. **1**

____

____

Total marks 6

[END OF QUESTION PAPER]

MARKS DO NOT WRITE IN THIS MARGIN

ADDITIONAL SPACE FOR ANSWERS

ADDITIONAL GRAPH PAPER FOR QUESTION 1(b)

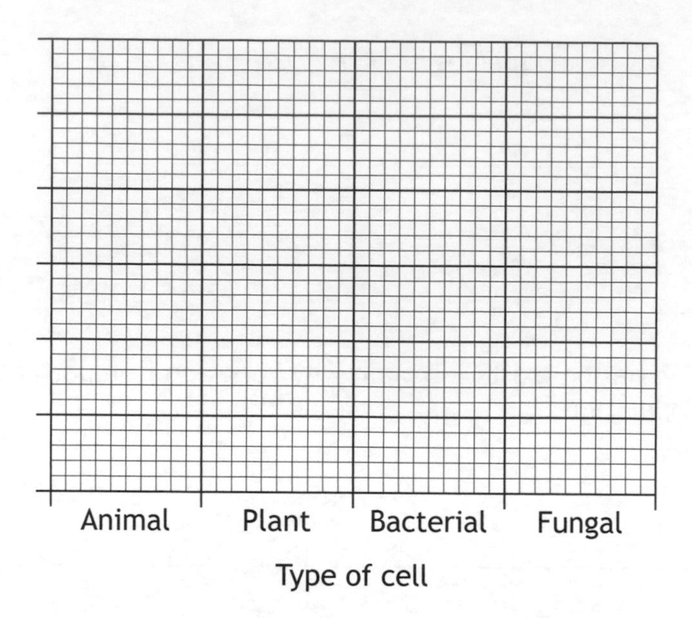

ADDITIONAL PIE CHART FOR QUESTION 11(b)

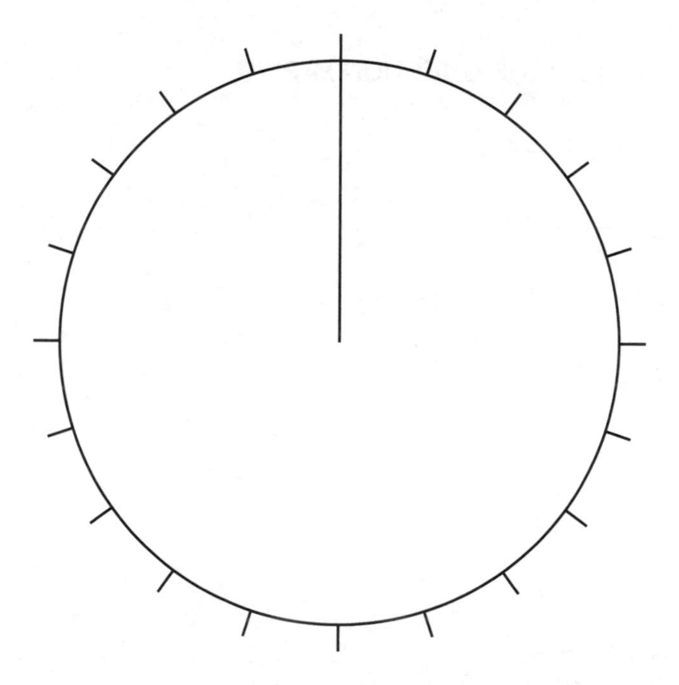

MARKS DO NOT WRITE IN THIS MARGIN

ADDITIONAL SPACE FOR ANSWERS AND ROUGH WORK

MARKS DO NOT WRITE IN THIS MARGIN

ADDITIONAL SPACE FOR ANSWERS AND ROUGH WORK

NATIONAL 5

2015

X707/75/02

Biology
Section 1—Questions

WEDNESDAY, 13 MAY

9:00 AM – 11:00 AM

Instructions for the completion of Section 1 are given on Page two of your question and answer booklet X707/75/01.

Record your answers on the answer grid on Page three of your question and answer booklet.

Before leaving the examination room you must give your question and answer booklet to the Invigilator; if you do not, you may lose all the marks for this paper.

SECTION 1

1. In the diagrams below, the circles represent molecules on either side of a cell membrane. In which of these diagrams would the molecules move into a cell by diffusion?

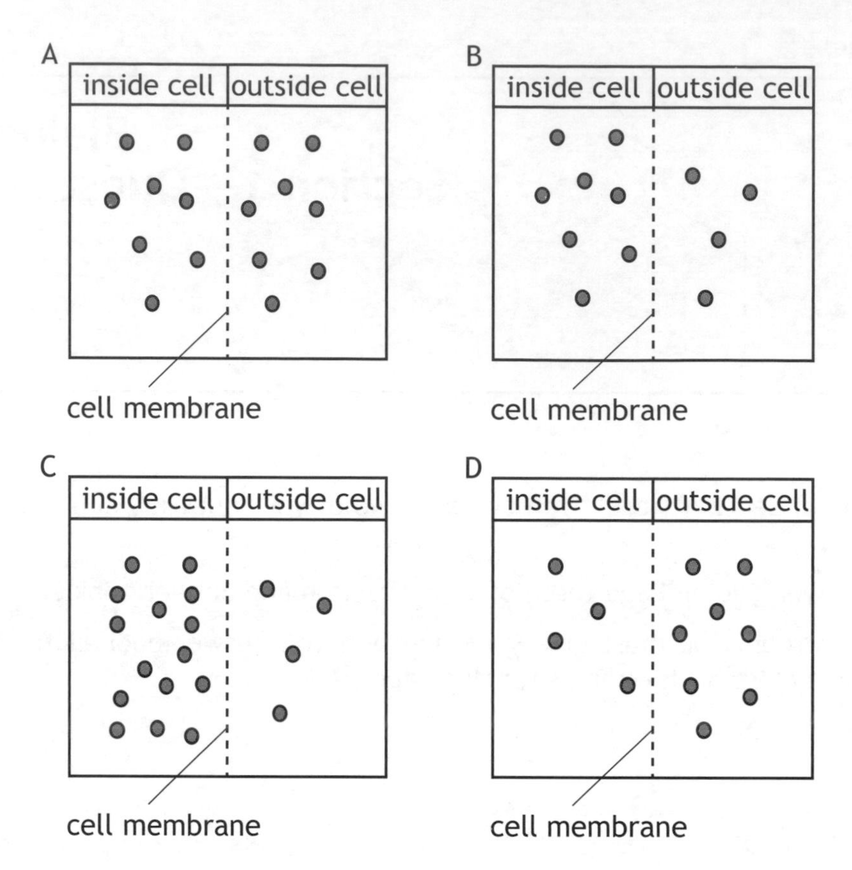

2. Which of the following does **not** involve mitosis?

   A Synthesis of proteins

   B Growth of tissue

   C Maintenance of the diploid chromosome complement

   D Repair of tissue

3. The graph below shows changes in the enzyme and substrate concentrations in a seed over a period of time.

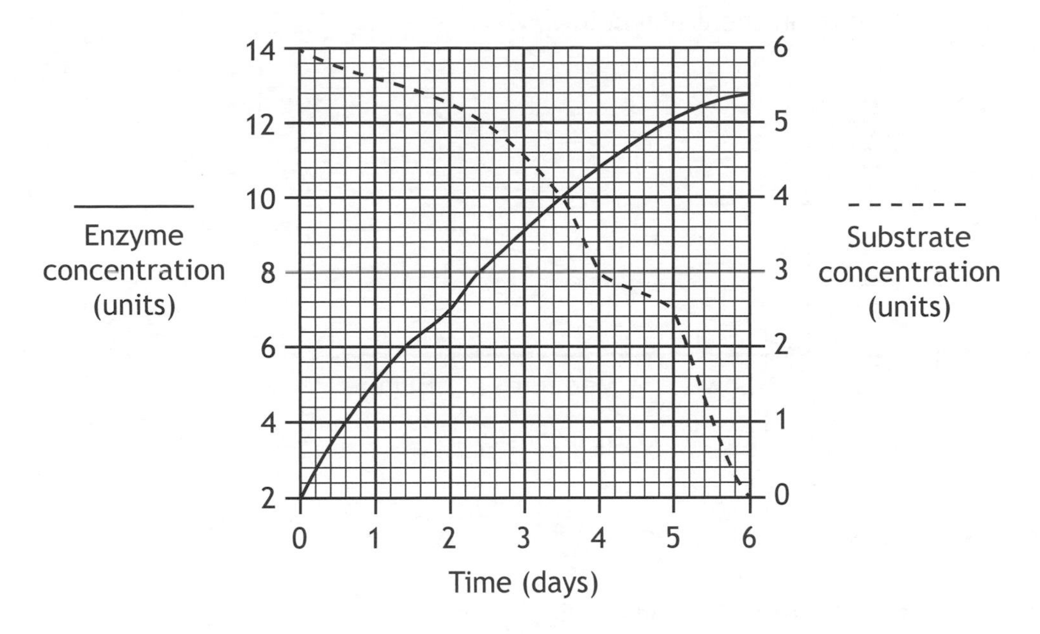

How many days does it take for the substrate concentration to decrease by 50%?

A 2

B 3

C 4

D 5

4. Some stages of genetic engineering are shown below.

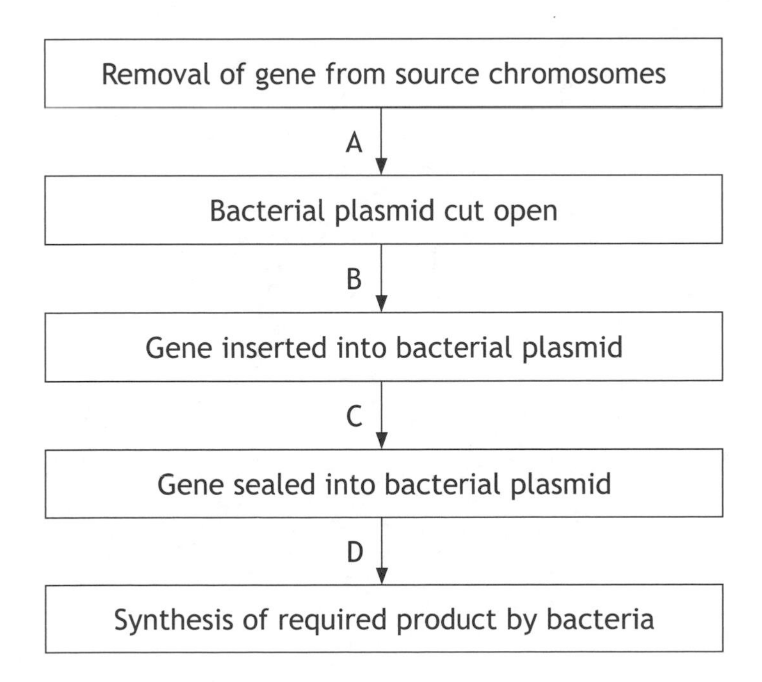

Which letter indicates the stage where the plasmid is inserted into a bacterial cell?

5. The effect of light intensity on the rate of photosynthesis was measured for two species of plants, L and M.

The results are shown in the graph below.

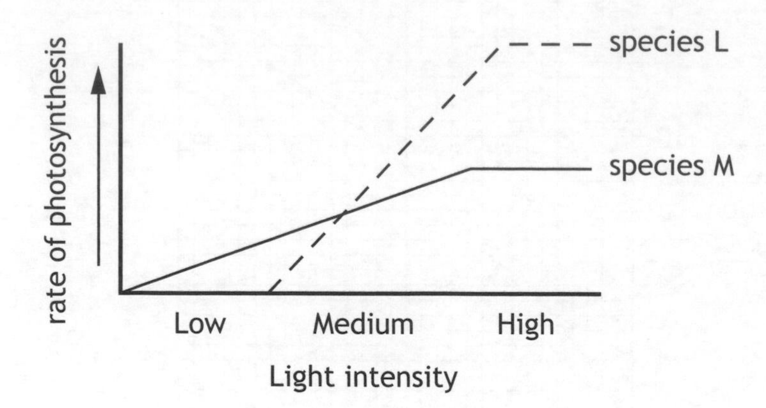

The rate of photosynthesis of species M is

A slower than L in low light intensities

B slower than L in high light intensities

C faster than L in medium light intensities

D faster than L in high light intensities.

6. The diagrams below show four different types of cell.

Which cell was produced by a meristem?

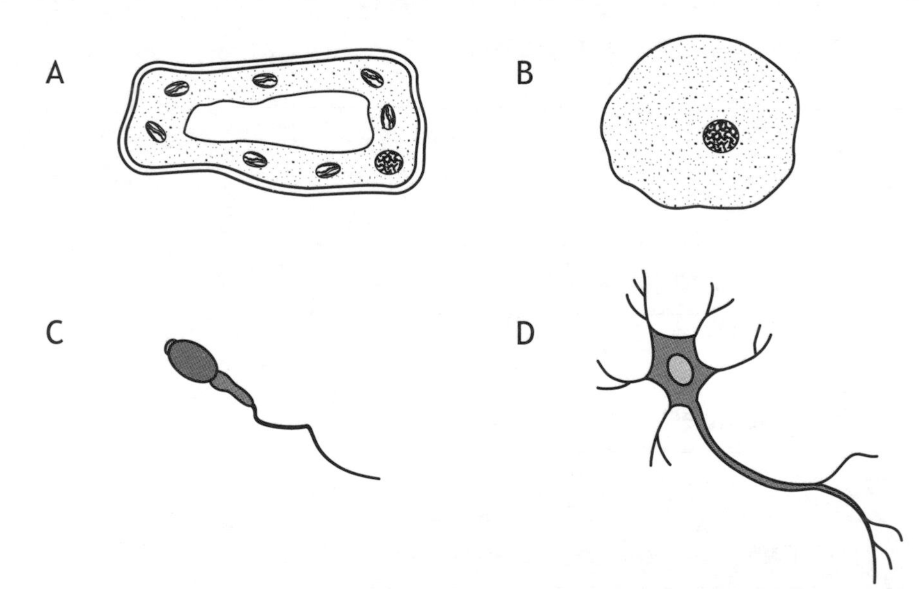

7. The diploid number of chromosomes in a cell from a kangaroo is 12.

Which line in the table below identifies the number of chromosomes for the cell type shown?

	Kangaroo Cell Type	*Number of chromosomes*
A	sperm	12
B	skin	6
C	nerve	6
D	zygote	12

8. The diagrams below show the same sections of matching chromosomes found in four flies, A, B, C and D.

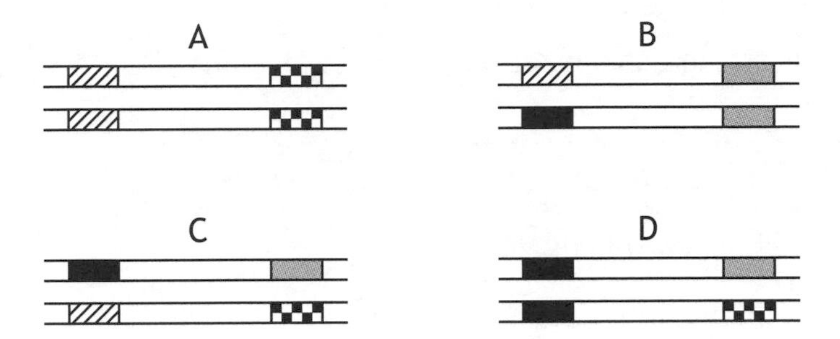

The alleles shown on the chromosomes can be identified using the following key.

- allele for striped body
- allele for unstriped body
- allele for normal antennae
- allele for abnormal antennae

Which fly is homozygous for body pattern and heterozygous for antennae type?

[Turn over

9. The diagram below shows an alveolus and an associated blood capillary.

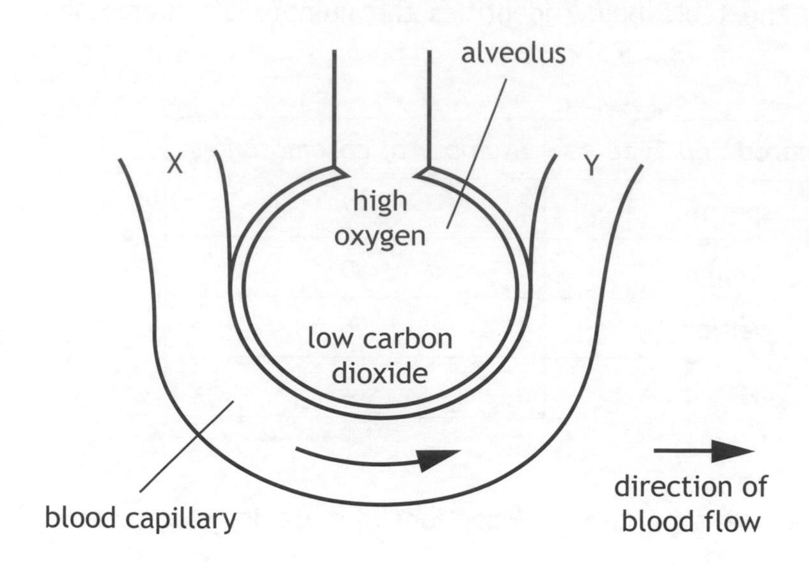

As blood flows from X to Y gases are exchanged with the alveolus.

Which line in the table below identifies the concentrations of gases at X and Y?

	Concentration at X	*Concentration at Y*
A	high oxygen	high carbon dioxide
B	low oxygen	high carbon dioxide
C	low oxygen	low carbon dioxide
D	high oxygen	low carbon dioxide

10. The following sequence shows part of the blood flow through the body.

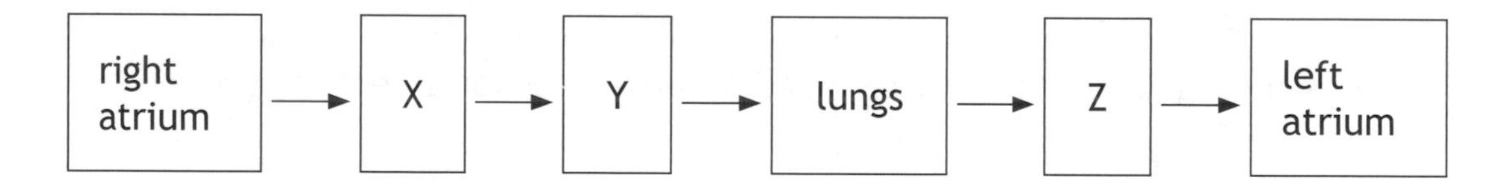

Which line in the table below identifies X, Y and Z?

	X	*Y*	*Z*
A	right ventricle	pulmonary vein	pulmonary artery
B	right ventricle	pulmonary artery	pulmonary vein
C	pulmonary vein	pulmonary artery	right ventricle
D	pulmonary artery	right ventricle	pulmonary vein

11. The graph below shows the relationship between the concentration of carbon dioxide and oxyhaemoglobin in the blood.

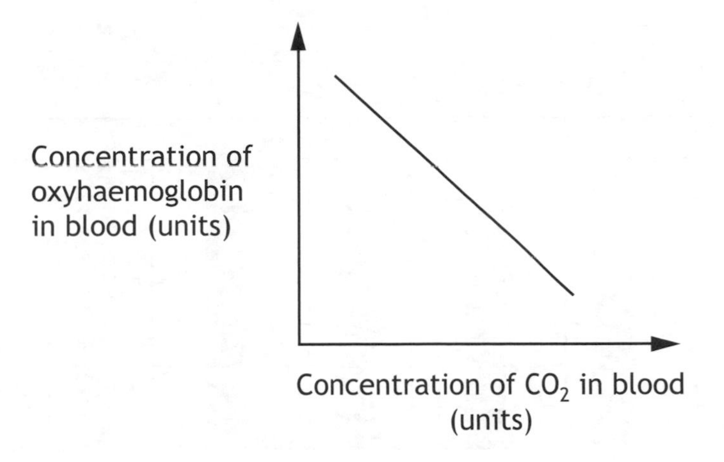

Which of the following statements describes this relationship?

A As the carbon dioxide concentration increases the concentration of oxyhaemoglobin decreases.

B As the carbon dioxide concentration decreases the concentration of oxyhaemoglobin decreases.

C As the carbon dioxide concentration increases the concentration of oxyhaemoglobin increases.

D Increasing carbon dioxide concentration has no effect upon the concentration of oxyhaemoglobin.

[Turn over

12. The chart below shows the percentage of men and women with obesity at different ages, in a population.

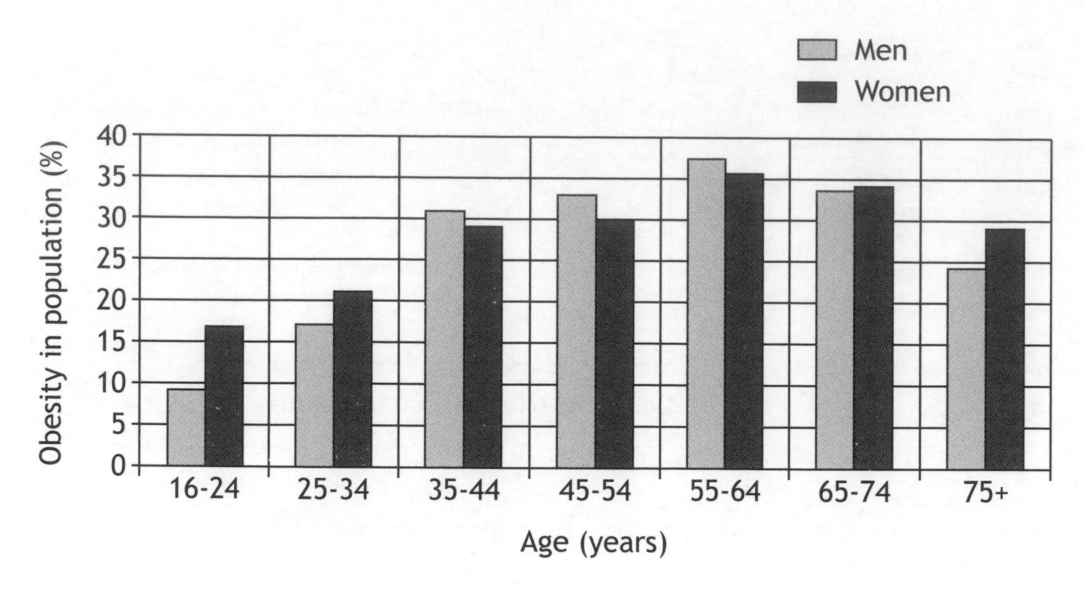

Which of the following statements is true?

A For each age group there is a higher percentage of obese men than obese women.

B For each age group there is a higher percentage of obese women than obese men.

C Obesity in men and women increases with age up to 64 years.

D Obesity in men and women decreases with age up to 64 years.

13. Which of the following statements best describes a biome?

A All the organisms in an area and their habitat.

B The role that an organism plays within a community.

C A living factor which affects biodiversity in an ecosystem.

D A region of our planet as distinguished by its climate, fauna and flora.

14. The size of a population of snails can be estimated using the following formula.

$$\text{Population} = \frac{\text{Number collected on 1st day} \times \text{Number collected on 2nd day}}{\text{Number of marked individuals found on 2nd day}}$$

A student investigated the population of snails in a garden. He collected 40 snails, marked their shells and released them. Next day, 35 snails were collected and 14 of these were found to be marked.

The snail population was estimated to be

A 16

B 100

C 560

D 1400.

15. Which of the following describes interspecific competition?

A Individuals of different species requiring different resources.

B Individuals of different species requiring similar resources.

C Individuals of the same species requiring different resources.

D Individuals of the same species requiring similar resources.

16. The diagram below represents four populations of animals P, Q, R and S and areas of interbreeding. Interbreeding takes place in the shaded areas.

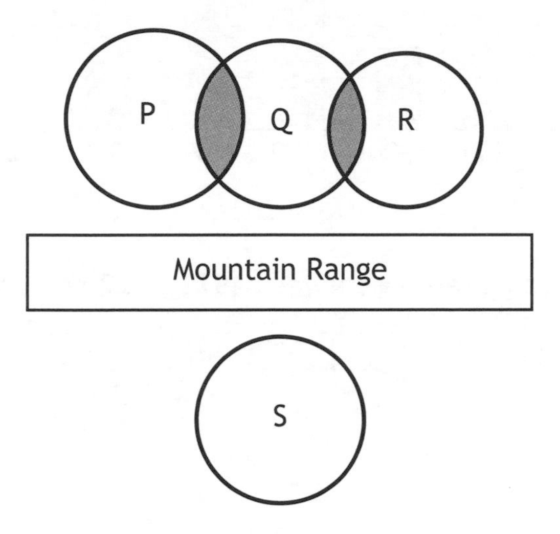

How many species may evolve over time?

A 1

B 2

C 3

D 4

17. Antibiotic resistance in bacteria is an example of evolution. Which of the following shows the sequence of events leading to this?

A Natural selection → mutation → use of antibiotic

B Mutation → natural selection → use of antibiotic

C Mutation → use of antibiotic → natural selection

D Natural selection → use of antibiotic → mutation

[Turn over

18. The graph below shows information about the growth of the human population.

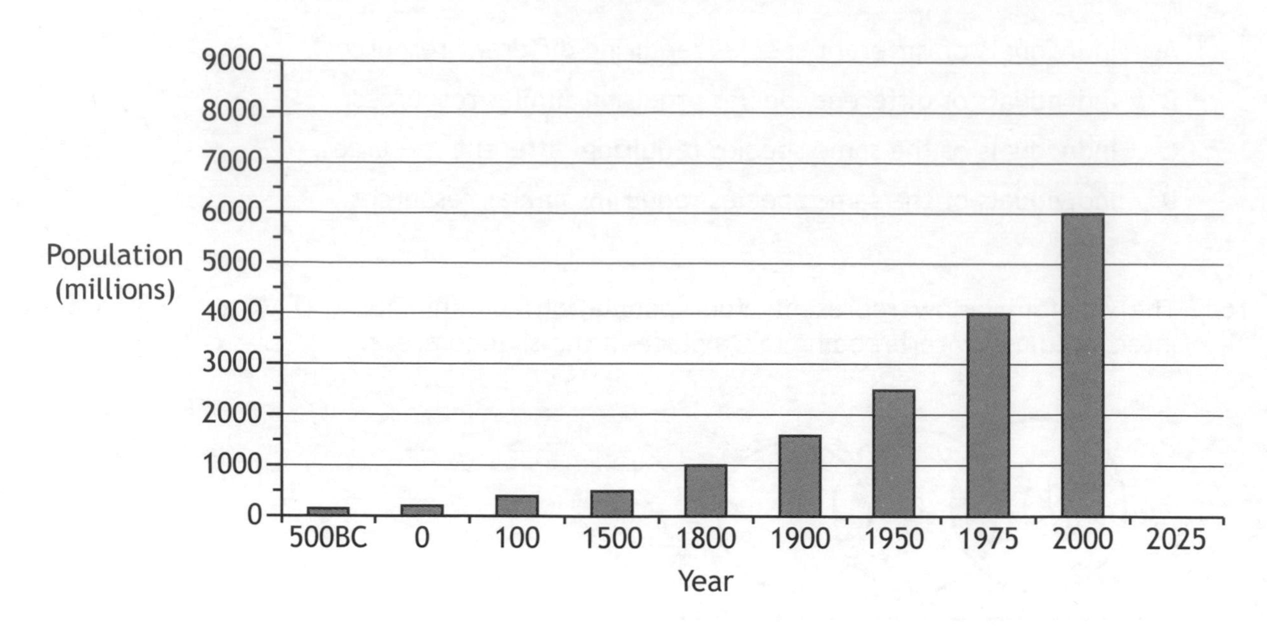

If the population continues to increase at the same rate as between 1975 and 2000, predict the population size in 2025.

A 7000

B 7500

C 8000

D 8500

19. DDT can be sprayed onto crops to kill insects. It can be washed off the crops by rainwater and flow into rivers where it accumulates in food chains.

A typical freshwater food chain and the concentration of DDT in each organism is shown below.

Food chain:	algae →	stickleback →	trout →	osprey
DDT concentration:	0·001	2·0	5·0	20·0

The percentage increase in DDT concentration between the trout and osprey is

A 15

B 100

C 300

D 400.

20. Which of the following statements describes the sequence of events when fertiliser leaches into a loch?

A Algal bloom develops ⟶ algae die ⟶ oxygen concentration increases

B Algal bloom develops ⟶ algae die ⟶ oxygen concentration decreases

C Oxygen concentration increases ⟶ algal bloom develops ⟶ algae die

D Algae die ⟶ oxygen concentration decreases ⟶ algal bloom develops

[END OF SECTION 1. NOW ATTEMPT THE QUESTIONS IN SECTION 2 OF YOUR QUESTION AND ANSWER BOOKLET]

[BLANK PAGE]

DO NOT WRITE ON THIS PAGE

FOR OFFICIAL USE

National Qualifications 2015

Mark

X707/75/01

Biology
Section 1—Answer Grid and Section 2

WEDNESDAY, 13 MAY

9:00 AM – 11:00 AM

Fill in these boxes and read what is printed below.

Full name of centre

Town

Forename(s)

Surname

Number of seat

Date of birth

Day Month Year

Scottish candidate number

Total marks — 80

SECTION 1 — 20 marks

Attempt ALL questions.

Instructions for the completion of Section 1 are given on Page two.

SECTION 2 — 60 marks

Attempt ALL questions in this section.

Write your answers clearly in the spaces provided in this booklet. Additional space for answers and rough work is provided at the end of this booklet. If you use this space you must clearly identify the question number you are attempting. Any rough work must be written in this booklet. You should score through your rough work when you have written your final copy.

Use **blue** or **black** ink.

Before leaving the examination room you must give this booklet to the Invigilator; if you do not, you may lose all the marks for this paper.

SECTION 1— 20 marks

The questions for Section 1 are contained in the question paper X707/75/02.
Read these and record your answers on the answer grid on *Page three* opposite.
Use **blue** or **black** ink. Do NOT use gel pens or pencil.

1. The answer to each question is **either** A, B, C or D. Decide what your answer is, then fill in the appropriate bubble (see sample question below).

2. There is **only one correct** answer to each question.

3. Any rough working should be done on the additional space for answers and rough work at the end of this booklet.

Sample Question

The thigh bone is called the

A humerus

B femur

C tibia

D fibula.

The correct answer is **B**—femur. The answer **B** bubble has been clearly filled in (see below).

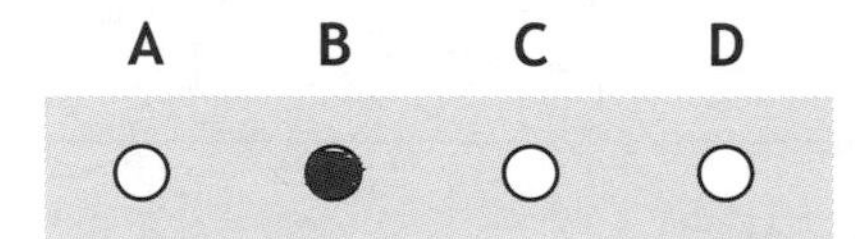

Changing an answer

If you decide to change your answer, cancel your first answer by putting a cross through it (see below) and fill in the answer you want. The answer below has been changed to **D**.

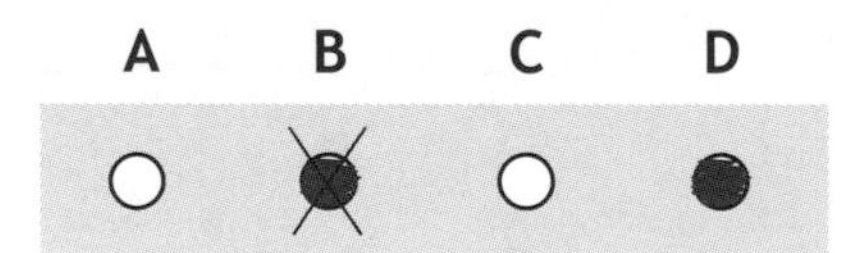

If you then decide to change back to an answer you have already scored out, put a tick (✓) to the **right** of the answer you want, as shown below:

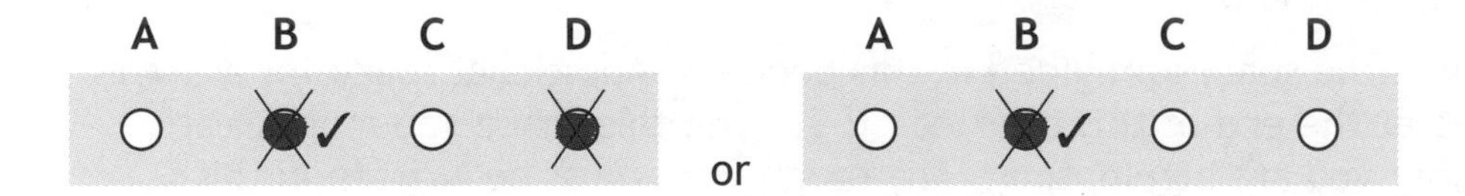

SECTION 1 — Answer Grid

	A	B	C	D
1	○	○	○	○
2	○	○	○	○
3	○	○	○	○
4	○	○	○	○
5	○	○	○	○
6	○	○	○	○
7	○	○	○	○
8	○	○	○	○
9	○	○	○	○
10	○	○	○	○
11	○	○	○	○
12	○	○	○	○
13	○	○	○	○
14	○	○	○	○
15	○	○	○	○
16	○	○	○	○
17	○	○	○	○
18	○	○	○	○
19	○	○	○	○
20	○	○	○	○

[BLANK PAGE]

DO NOT WRITE ON THIS PAGE

MARKS | DO NOT WRITE IN THIS MARGIN

SECTION 2 — 60 marks

Attempt ALL questions

1. (a) The diagram below represents a cell in an early stage of mitosis.

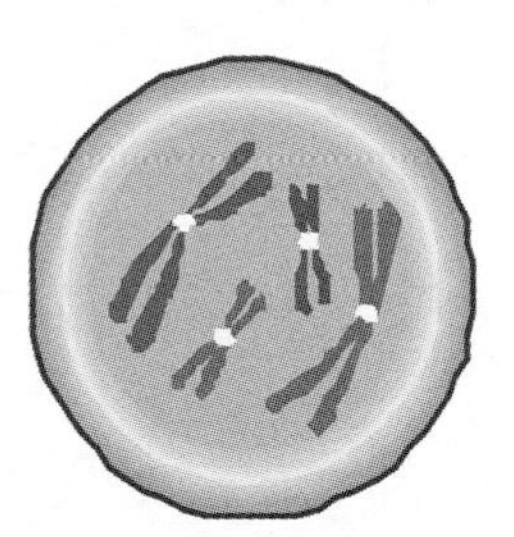

(i) State the number of chromosomes present in this cell. **1**

____

(ii) State how many chromosomes will be present in each of the two cells produced by the process. **1**

____

(b) Name a site of mitosis in plants. **1**

____

____

[Turn over

MARKS | DO NOT WRITE IN THIS MARGIN

2. (a) Shells can be removed from eggs by dissolving them in vinegar for 2–3 days. The egg contents remain inside a thin membrane.

In an investigation the shells from two eggs were removed. The eggs were then weighed and placed in beakers as shown below.

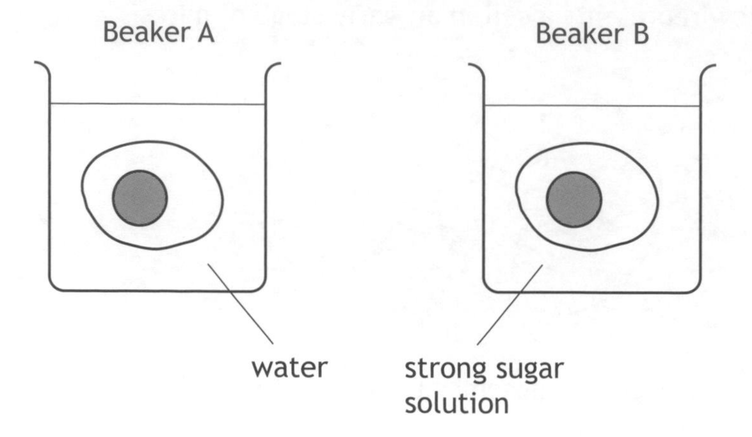

After 2 hours the eggs were removed from the beakers, blotted dry and reweighed. The results are shown in the following table.

Beaker	*Mass at start (g)*	*Mass after 2 hours (g)*	*Percentage change in mass*
A	54·0	67·5	
B	52·1	47·8	−8·2

(i) Complete the table by calculating the percentage change in mass for beaker A. **1**

Space for calculation

(ii) Suggest why the eggs were blotted dry before being reweighed. **1**

MARKS | DO NOT WRITE IN THIS MARGIN

2. (a) (continued)

(iii) Choose either beaker A or B and explain how osmosis caused the change in mass of the eggs in that beaker. **2**

Beaker ____________

Explanation ____

____

____

(b) The movement of molecules in or out of cells can be by passive or active transport.

Describe **one** difference between passive and active transport. **1**

____

____

[Turn over

MARKS | DO NOT WRITE IN THIS MARGIN

3. (a) DNA is a double stranded molecule. The following diagram shows part of one strand. Complete the diagram to show the complementary strand.

DNA Strand

A T G C G A T G C G C T G T C

Complementary DNA Strand

1

(b) (i) DNA contains genetic material which controls the synthesis of chemicals made from amino acids.

Name the type of chemicals synthesised. 1

____

(ii) The diagram below shows an example of one of these chemicals being synthesised.

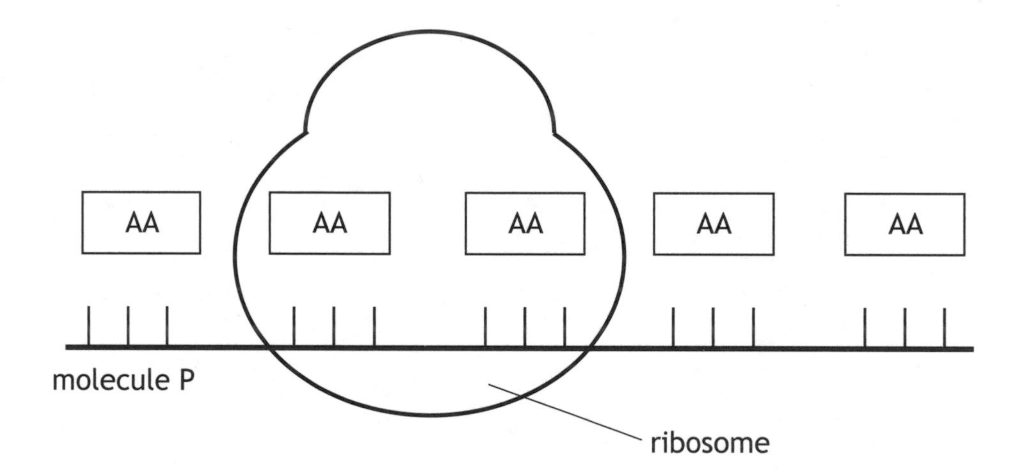

Name molecule P and describe how it determines the sequence of amino acids, represented by AA, as shown in the diagram. 2

Molecule P ______________________________

Description ______________________________

____

(iii) Name the part of the cell where molecule P was made. 1

____

MARKS | DO NOT WRITE IN THIS MARGIN

4. Photosynthesis is a two stage process used by green plants to produce food.

(a) The diagram below represents a summary of the first stage of photosynthesis.

Complete the diagram by filling in the three boxes, selecting terms from the list in the box below. **3**

ATP	**carbon dioxide**	**carbon fixation**	
sugar	**hydrogen**	**oxygen**	**light reactions**

Name of the first stage

____

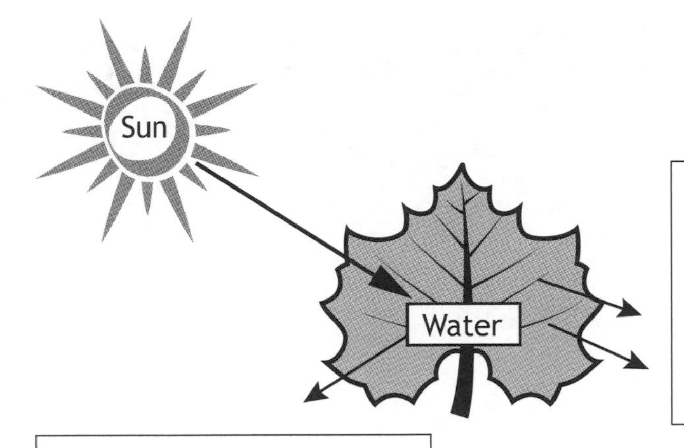

Two products used in second stage.

1. ____________________

2. ____________________

Diffuses out of the leaf

____

(b) Describe the second stage of photosynthesis. **2**

____

____

____

____

[Turn over

MARKS | DO NOT WRITE IN THIS MARGIN

5. (a) Cellular processes occur in different parts of the cell.

Name the energy producing process which starts in the cytoplasm and is completed in the mitochondria. **1**

____

(b) As a result of the complete breakdown of a number of glucose molecules, 114 molecules of ATP were produced.

State the number of glucose molecules which were broken down to achieve this. **1**

Space for calculation

___________ Glucose molecules

(c) Explain why a sperm cell contains more mitochondria than a skin cell. **1**

____

____

____

MARKS | DO NOT WRITE IN THIS MARGIN

6. The diagram below shows the neurons involved in a reflex action. Neurons J, K and L form a reflex arc.

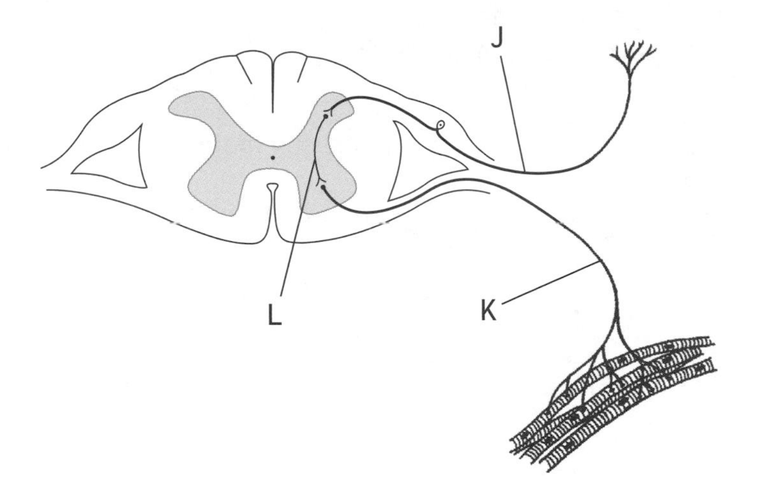

(a) Describe how information is passed along a neuron. **1**

____

(b) Select **one** of the neurons shown in the diagram and tick (✓) the appropriate box below.

Name that type of neuron and describe its particular function. **2**

J ☐ K  L 

Name ____

Function ______________________________________

____

____

(c) During a reflex action, the speed at which the information flows was measured to be 90 metres per second.

Calculate how long it would take for the information to complete a reflex arc which was 0·9 m in length. **1**

Space for calculation

______________ seconds

MARKS | DO NOT WRITE IN THIS MARGIN

7. (a) One type of deafness in humans is caused by a single gene.

The diagram below shows the pattern of inheritance in one family.

H represents the hearing form of the gene.
h represents the non-hearing form of the gene.

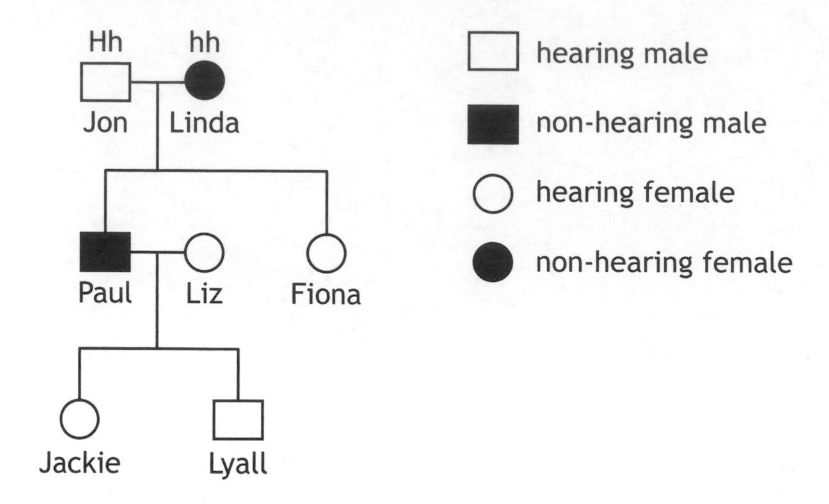

(i) Using Jon as an example, explain how it is known that the hearing form of the gene is dominant. **1**

(ii) Use information in the family tree to complete the following table to show the genotype and phenotype of each individual. **2**

Individual	*Genotype*	*Phenotype*
Paul		
Lyall		

(iii) Fiona has a child with a man who has the same genotype as her. State the chance of their child being able to hear. **1**

Space for calculation

MARKS | DO NOT WRITE IN THIS MARGIN

7. **(continued)**

(b) Most features of an individual's phenotype are controlled by more than one gene.

Name this type of inheritance. **1**

____

[Turn over

MARKS DO NOT WRITE IN THIS MARGIN

8. (a) An experiment was set up to find out the optimum temperature for the growth of tomatoes in a glasshouse. The following table gives the results of this experiment.

Temperature (°C)	*Fresh mass of tomatoes* (g/plant)	*Dry mass of tomatoes* (g/plant)
14	1000	50
18	8300	415
22	9000	450
26	2200	110
32	1600	80

(i) On the grid below, complete the vertical axis and plot a line graph to show the effect of temperature on the dry mass of tomatoes. **2**

(Additional graph paper, if required, can be found on *Page twenty-three*)

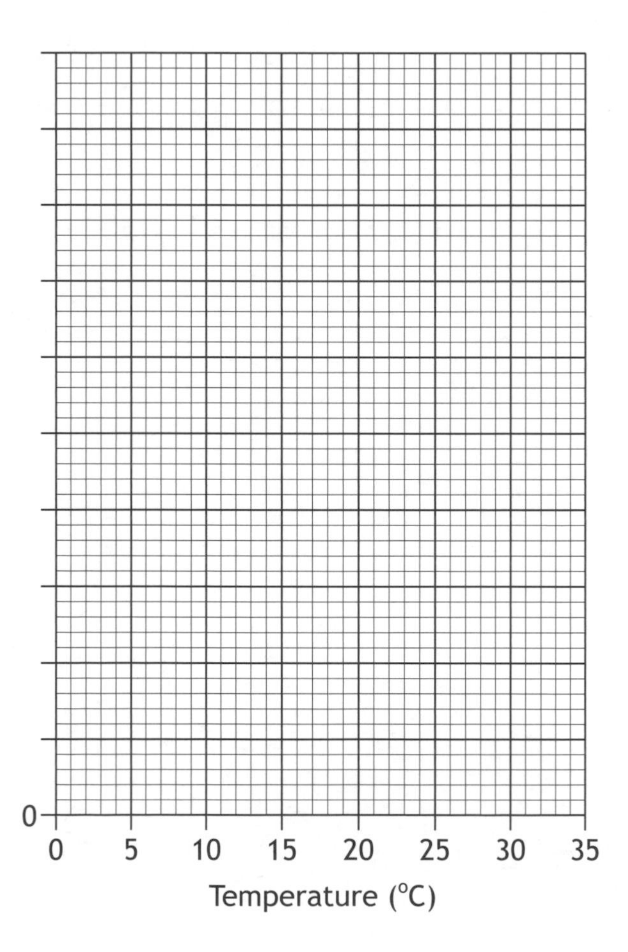

MARKS | DO NOT WRITE IN THIS MARGIN

8. (a) (continued)

(ii) Above 26 °C the drop in the fresh mass of tomatoes continues at a steady rate.

Using the information in the table, predict the fresh mass of tomatoes which will be produced at 35 °C. **1**

Space for calculation

_______________ g/plant

(b) The diagram below shows three parts of a plant.

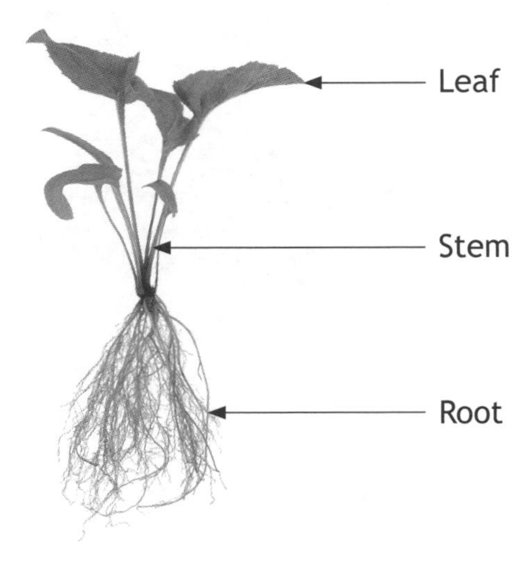

Describe the structures and processes involved as water moves through the plant from the soil to the air. **3**

____

____

____

____

____

____

____

____

____

____

MARKS | DO NOT WRITE IN THIS MARGIN

9. The diagrams below represent part of the human breathing system.

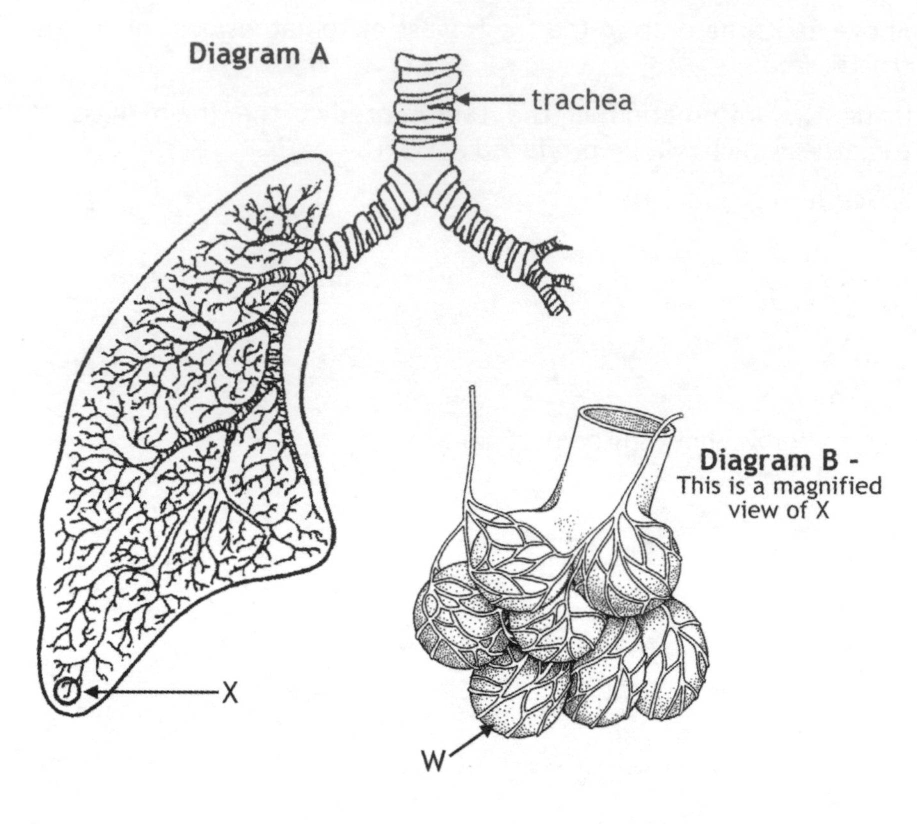

(a) (i) Name the structure labelled W. **1**

____

(ii) Describe **two** features of these structures which improve the efficiency of gas exchange. **2**

1 ____

2 ____

(b) Mucus and cilia are found in the trachea.

Describe how the mucus and cilia work together to help prevent bacteria getting into the lungs. **2**

____

____

____

____

MARKS DO NOT WRITE IN THIS MARGIN

10. Nitrogen is an important element in living organisms. The diagram below shows stages in the transfer of nitrogen in an ecosystem.

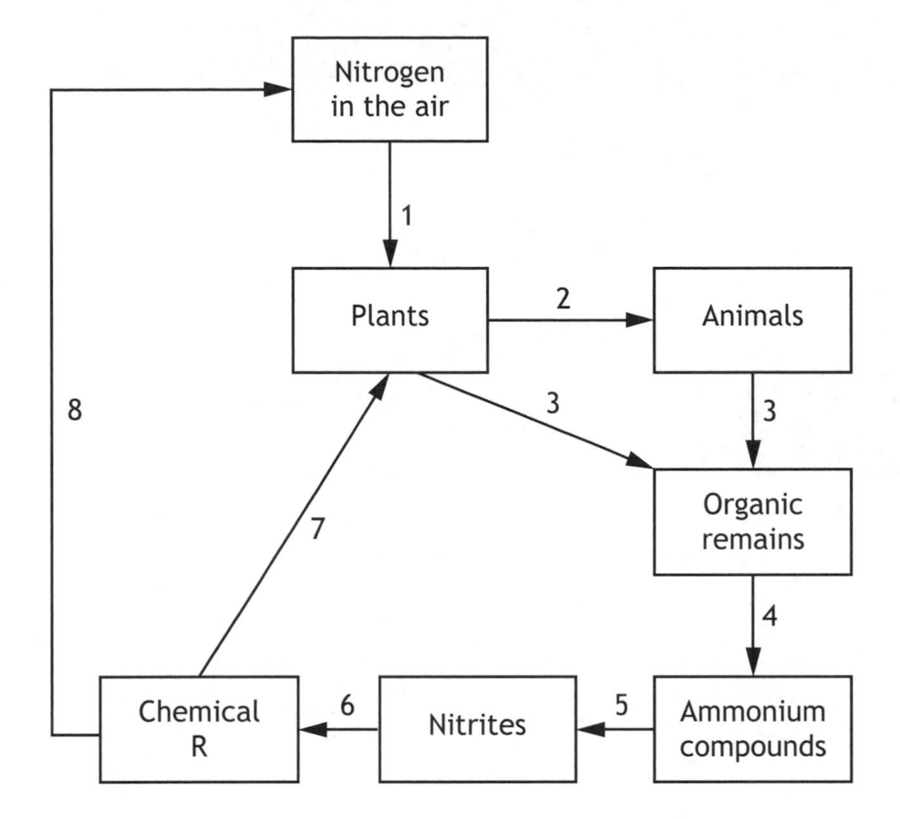

(a) The numbers in the diagram above represent stages in the transfer of nitrogen in an ecosystem.

Select the correct number(s) to complete the table below to identify the named stages. **2**

Stage	*Number*
Death and decay	
Denitrification	

(b) Nitrogen fixing bacteria are involved in stage 1.

State **one** place where these microorganisms can be found. **1**

____

(c) Identify chemical R and explain its importance to plants. **2**

Chemical R ____________________________

Importance to plants ________________________

____

[Turn over

MARKS | DO NOT WRITE IN THIS MARGIN

11. A river was sampled at five sites as shown in the diagram below.

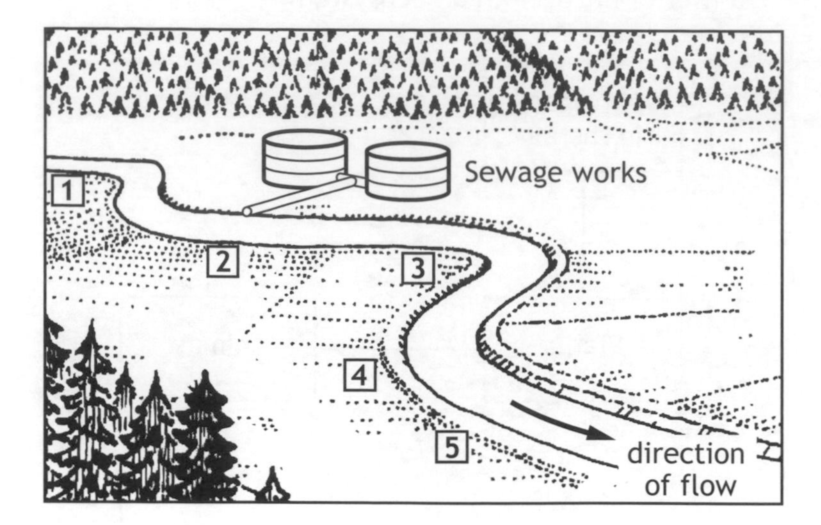

The following tables show the results of analysing the samples at each site.

Table 1

Site	*Oxygen levels* (Units)	*Number of bacteria per 100ml*
1	1·2	500
2	0·04	150 000
3	0·40	12 680
4	0·54	3 400
5	1·12	1 250

Table 2

Organism Present	*Site 1*	*Site 2*	*Site 3*	*Site 4*	*Site 5*
Mayfly nymphs	23	0	0	0	8
Stonefly nymphs	42	0	0	0	21
Caddis fly larvae	18	0	0	10	15
Fresh water shrimp	2	0	0	1	1
Blood worms	1	5	24	7	1
Sludge worms	1	67	43	9	0

MARKS | DO NOT WRITE IN THIS MARGIN

11. (continued)

(a) (i) Using data from Table 1, describe the relationship between the number of bacteria and the oxygen level in the water. **1**

(ii) Methylene blue is a chemical which can be used to compare oxygen levels in the water. The lower the oxygen level, the faster methylene blue changes from blue to colourless.

A sample of water from each of the five sites was tested.

Predict which sample would lose its blue colour fastest. **1**

Sample from site number ________________

(b) Use data from Tables 1 and 2 to answer the following questions.

(i) State which of the organisms in the samples would be found in areas of high oxygen content. **1**

(ii) Sewage in the river is a form of water pollution.

Describe the effect this pollution has on the number of different types of organisms in this river. **1**

(c) Some species are known as indicator species.

Explain what is meant by indicator species. **1**

[Turn over

MARKS DO NOT WRITE IN THIS MARGIN

12. Ivy is a climbing plant which produces stems that grow vertically up trees and walls. It can also produce horizontal stems allowing the ivy to spread out along the ground.

Variation is shown in the width of the leaves of the ivy plant.

A group of students carried out an investigation to find out if the difference in leaf width is linked to the height of the leaves from the ground.

Five leaves were collected from a horizontal stem and another five from a vertical stem. The widths of the leaves were measured and the results are shown in the table below.

leaf	*Leaf width* (mm)	
	Horizontal stem	*Vertical stem*
1	52	32
2	60	34
3	56	35
4	50	44
5	52	35
average	54	

(a) Complete the table by calculating the average width of the leaves from the vertical stem. **1**

Space for calculation

MARKS | DO NOT WRITE IN THIS MARGIN

12. (continued)

(b) State the type of variation shown by leaf width. **1**

____

(c) The results show that leaves from a horizontal stem are bigger than leaves from a vertical stem.

Give a reason why these results might not be reliable. **1**

____

____

(d) To make the investigation valid, all leaves were taken from the same plant.

Explain why this was necessary. **1**

____

____

(e) The students wanted to find out what abiotic factors may have affected the width of the leaves from that plant.

Suggest **one** abiotic factor which they could have investigated. **1**

____

[Turn over

MARKS | DO NOT WRITE IN THIS MARGIN

13. Researchers have discovered an advantageous genetic mutation that causes high bone density in humans.

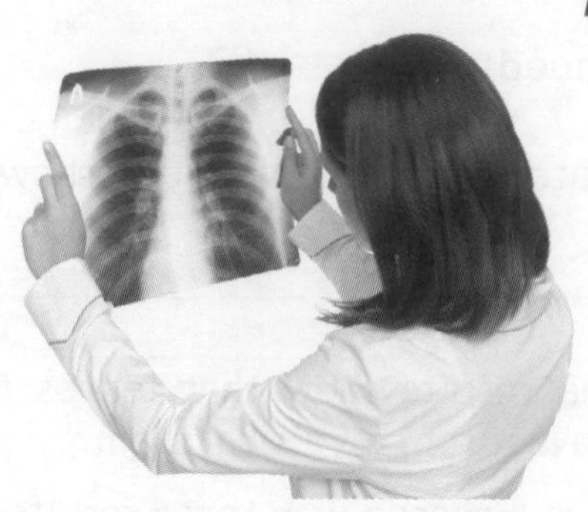

One man in the USA was discovered to possess this mutation after he walked away without injury from a serious car crash. Further studies have found several members of the same extended family with this mutation.

20 members of the family provided blood samples for DNA and biochemical testing. 7 of them were found to have high bone density. The same tests were performed on another group of 20 unrelated individuals with normal bone density.

The location of the gene mutation was able to be identified and it is hoped that the findings will help in developing medications to increase bone density for the treatment of conditions such as osteoporosis.

(a) (i) Calculate the percentage of the family who did **not** have the mutation for high bone density. **1**

Space for calculation

_______________ %

(ii) Explain why the biochemical tests were also performed on the 20 individuals with normal bone density. **1**

____

____

(b) Name **one** factor which can increase the rate of mutation. **1**

____

(c) Mutations are the only source of new alleles.

Explain why it is important that new alleles arise in a species. **1**

____

____

____

[END OF QUESTION PAPER]

MARKS DO NOT WRITE IN THIS MARGIN

ADDITIONAL SPACE FOR ANSWERS AND ROUGH WORK

ADDITIONAL GRAPH PAPER FOR QUESTION 8(a)(i)

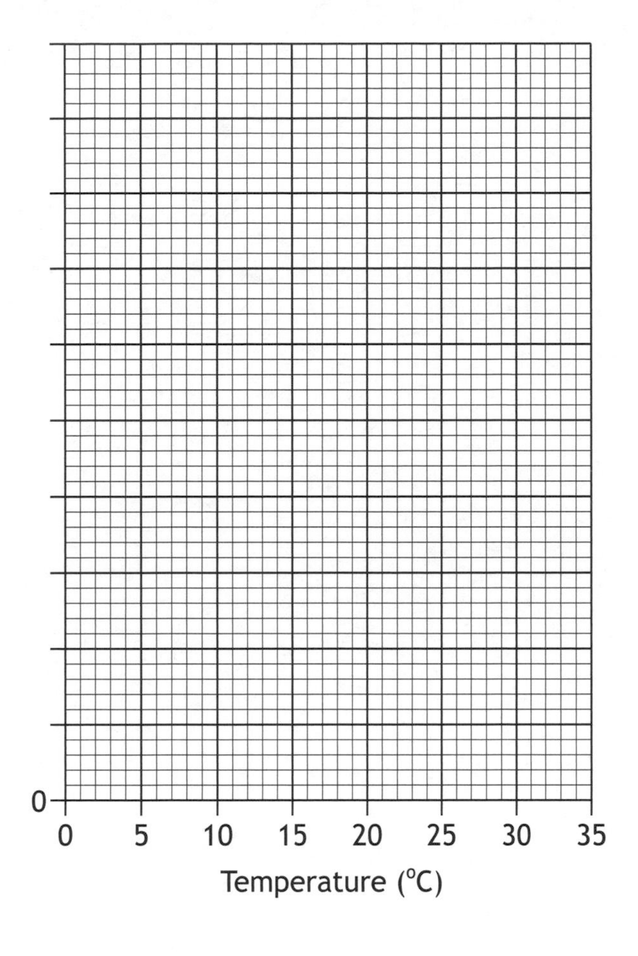

MARKS | DO NOT WRITE IN THIS MARGIN

ADDITIONAL SPACE FOR ANSWERS AND ROUGH WORK

MARKS | DO NOT WRITE IN THIS MARGIN

ADDITIONAL SPACE FOR ANSWERS AND ROUGH WORK

[BLANK PAGE]

DO NOT WRITE ON THIS PAGE

NATIONAL 5

2016

X707/75/02

Biology
Section 1—Questions

MONDAY, 9 MAY

1:00 PM – 3:00 PM

Instructions for the completion of Section 1 are given on *Page two* of your question and answer booklet X707/75/01.

Record your answers on the answer grid on *Page three* of your question and answer booklet

Before leaving the examination room you must give your question and answer booklet to the Invigilator; if you do not, you may lose all the marks for this paper.

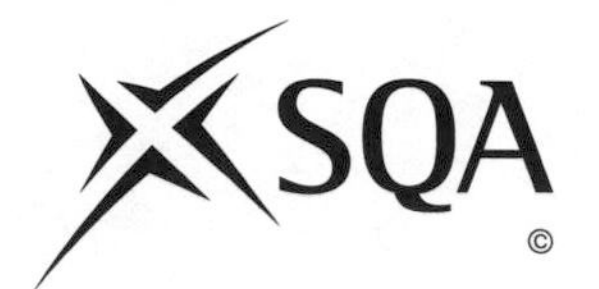

SECTION 1

1. The diagram below shows parts of a plant cell.

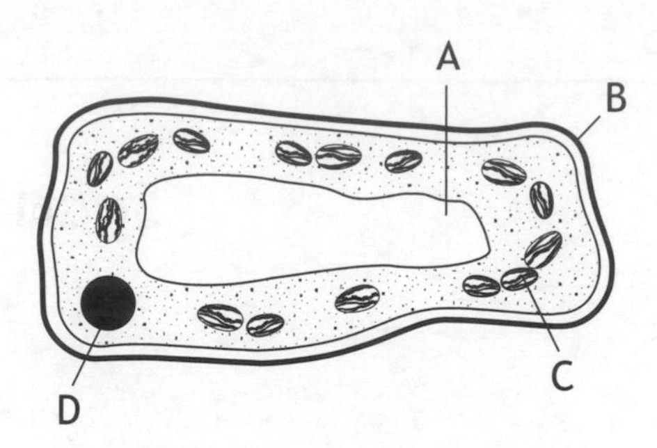

Which part of this cell is composed of cellulose?

2. Four cylinders of potato tissue were weighed and each was placed into a salt solution of a different concentration.

The cylinders were reweighed after one hour and the results are shown below.

Salt Solution	*Initial mass of potato cylinder* (g)	*Final mass of potato cylinder* (g)
A	10·0	7·0
B	10·0	9·4
C	10·0	11·2
D	10·0	12·6

In which salt solution would most potato cells be plasmolysed?

3. The diagram below shows the percentage of cells dividing in four areas of an onion root.

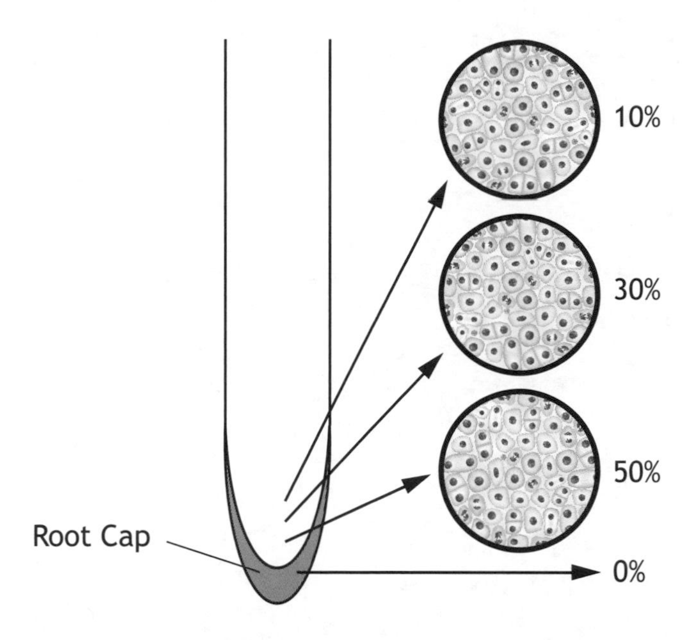

Which graph represents the number of cells dividing in this root?

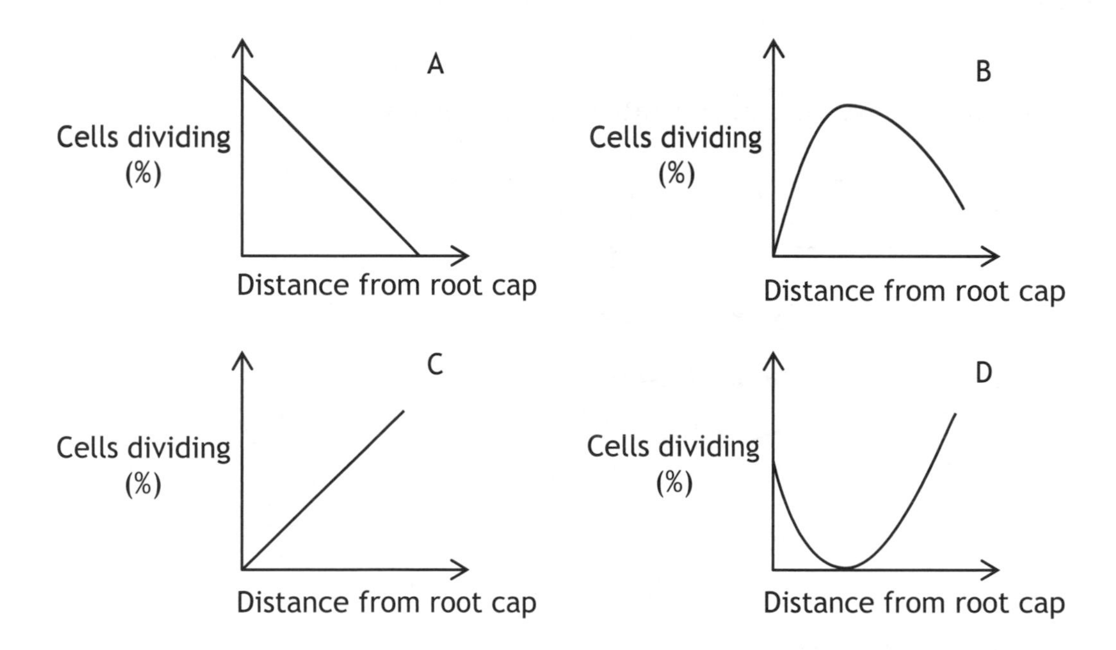

[Turn over

4. Which of the following shows the correct DNA base pairing?

A
A - C
C - G
G - C
T - A

B
A - T
C - G
G - T
T - A

C
A - G
C - G
G - A
T - A

D
A - T
C - G
G - C
T - A

5. Hormones are composed of

A glycerol
B glucose
C protein
D starch.

6. The diagram below shows the carbon fixation stage of photosynthesis.

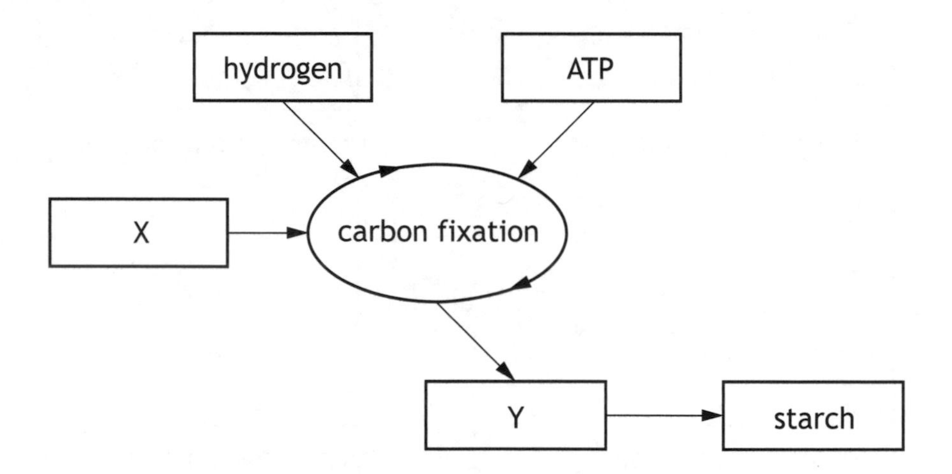

Which row in the table below identifies X and Y?

	X	Y
A	Sugar	Oxygen
B	Water	Carbon dioxide
C	Carbon dioxide	Sugar
D	Water	Oxygen

7. An investigation was carried out to compare the rate of oxygen gas production by two different species of water plant, S and T.

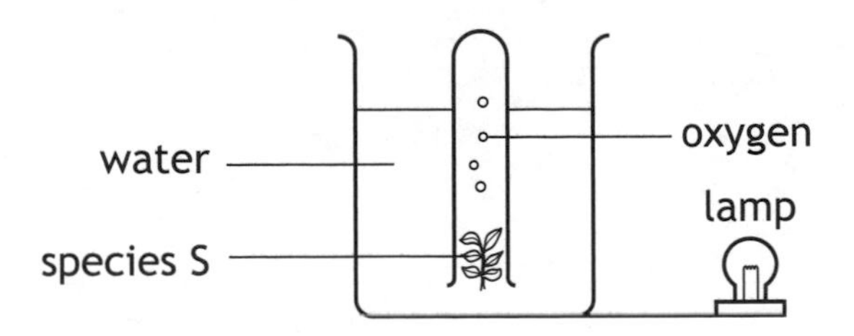

Which diagram below shows the set-up for species T, that would allow a valid comparison in the rate of oxygen production of the two species?

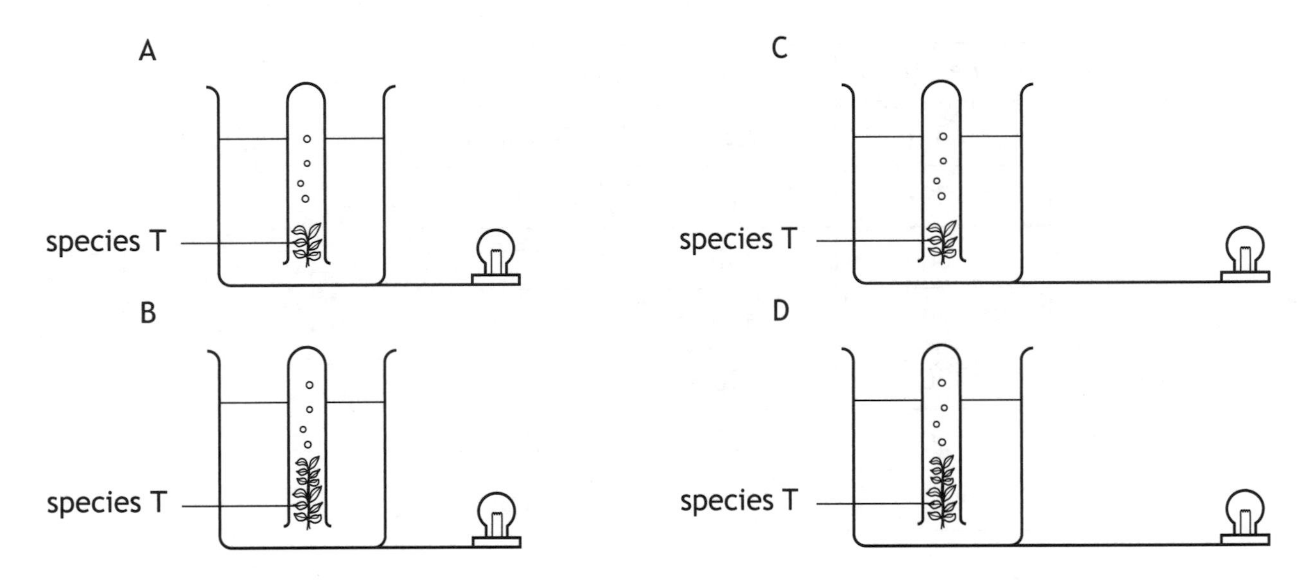

8. Each skin cell in a mouse has 40 chromosomes. How many chromosomes were present in each cell after dividing four times during cell culture?

A 10

B 20

C 40

D 160

[Turn over

9. Specialisation of cells in animals leads to the formation of

A tissues and organs

B meristems and organs

C stem cells and tissues

D stem cells and meristems.

10. The table below shows the blood glucose levels of two people after eating the same meal.

The normal range of blood glucose levels is 82–110 mg/dL.

Time after eating meal (min)	*Blood glucose levels* (mg/dL)	
	Person A	*Person B*
30	120	140
60	140	170
90	110	190
120	90	180
150	85	170
180	90	160

Using the information given, which of the following statements is correct?

A Person A always stayed within the normal range.

B Person B was outwith the normal range 180 minutes after eating.

C Person B had a level twice as high as that of person A 180 minutes after eating.

D Person A and person B both had their highest levels 90 minutes after eating.

11. The diagram below shows the structure of a flower.

Where are the male gametes produced?

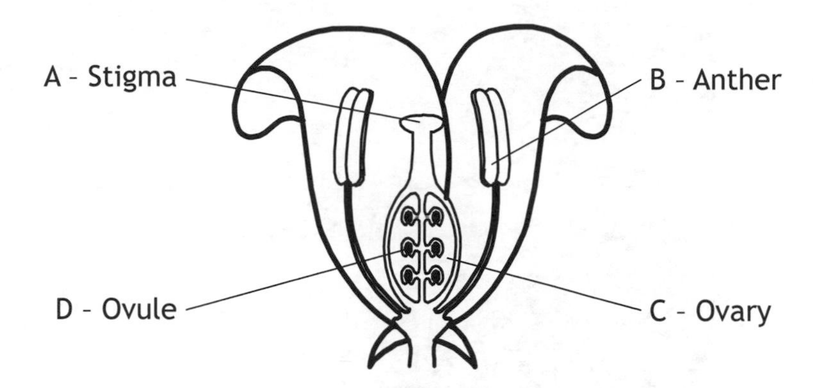

12. Most features of an individual phenotype are

A controlled by a single gene and show continuous variation

B controlled by a single gene and show discrete variation

C polygenic and show continuous variation

D polygenic and show discrete variation.

13. The following diagram shows the inheritance of coat colour in guinea pigs.

P Phenotype	Black guinea pig	X	White guinea pig
P Genotype:	BB		bb
F1 Genotype:		Bb	
F2 Genotypes:		BB and Bb and bb	

Which of the following generations contain heterozygous individuals?

A P and F1

B P and F2

C F1 and F2

D P, F1 and F2

[Turn over

14. The diagram below shows the heart and associated blood vessels.

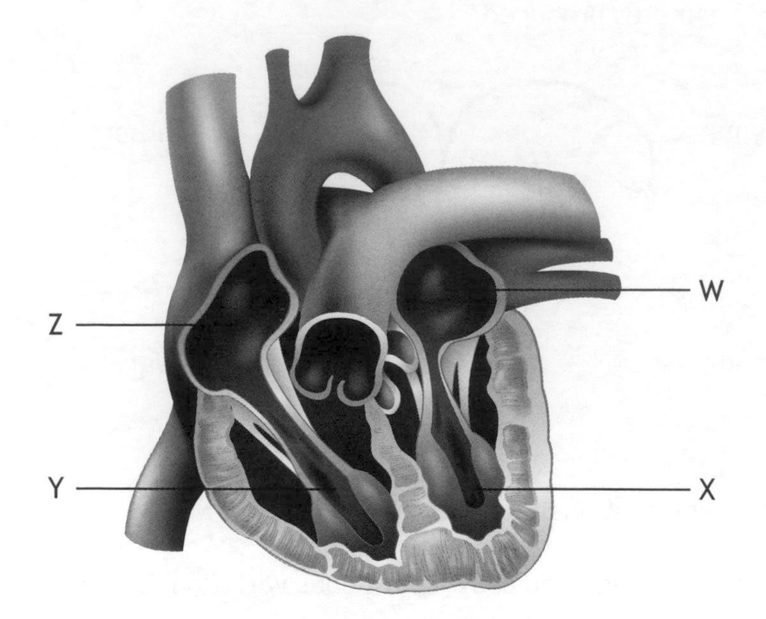

Which of the following statements is correct?

A W is the left atrium which receives blood from the body.

B X is the left ventricle which pumps blood to the body.

C Y is the right atrium which receives blood from the lungs.

D Z is the right ventricle which pumps blood to the lungs.

15. Which of the following statements best describes a niche?

A A living factor which affects biodiversity in an ecosystem.

B A region of our planet as distinguished by its climate, fauna and flora.

C All the organisms in an area and their habitat.

D The role that an organism plays within a community.

16. An ecosystem receives 6 000 000 units of energy from the sun.

Of this energy, 95% is **not** used in photosynthesis.

The amount of energy captured by the producers in this ecosystem is

A 30 000

B 300 000

C 570 000

D 5 700 000.

17. The graph below shows changes in the population of red and grey squirrels in an area of woodland over a 10 year period.

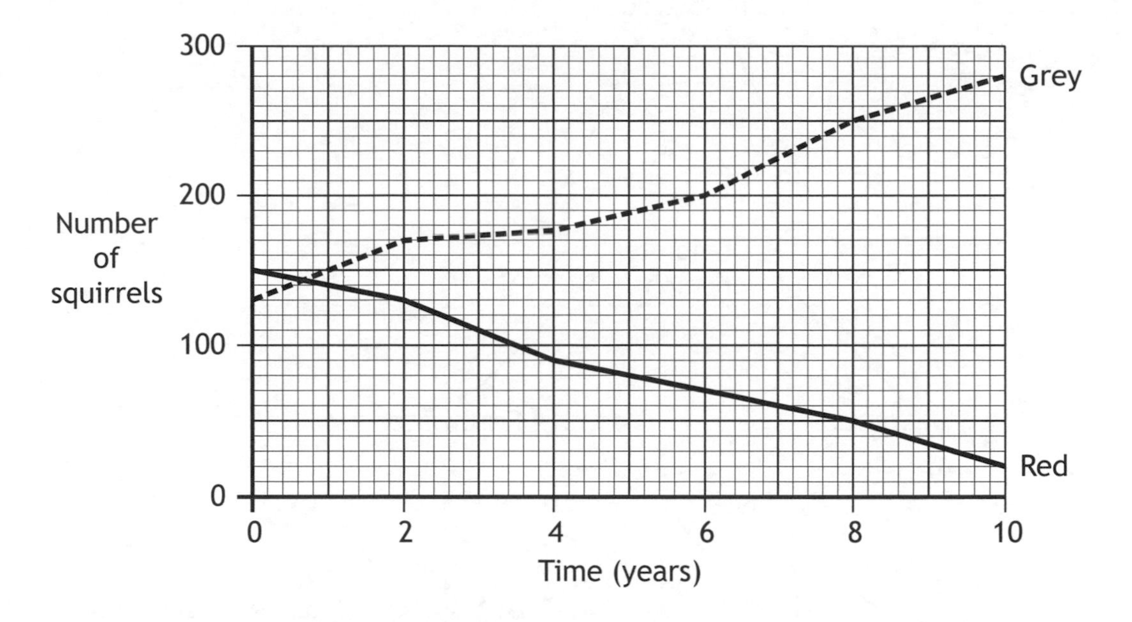

Which of the following conclusions can be drawn from the graph?

A The total number of squirrels decreased over 10 years.

B The population of red squirrels showed a greater change than the grey squirrels.

C The population of grey squirrels showed a greater change than the red squirrels.

D After 8 years there were 4 times as many grey squirrels as red squirrels.

18. Which of the following is a source of new alleles in a population?

A Mutation

B Isolation

C Natural selection

D Environmental conditions

19. Indicator species can provide information about

A numbers of organisms in a lake

B numbers of predators in a woodland

C levels of light in an ecosystem

D levels of pollution in a river.

[Turn over

20. The diagram below represents a freshwater food web.

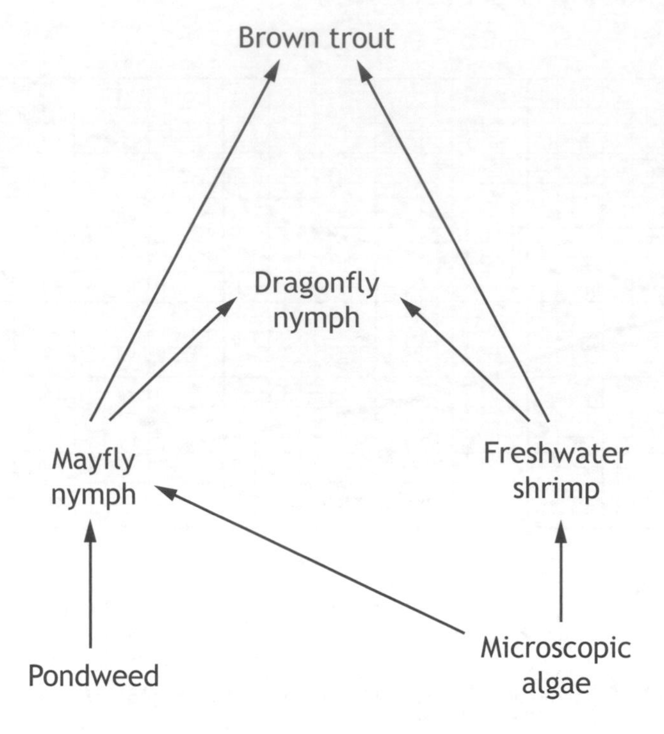

The number of freshwater shrimps was found to have decreased dramatically.

Predict the effect this will have on the numbers of dragonfly nymphs and microscopic algae.

A Both populations would decrease.

B Both populations would increase.

C Microscopic algae would decrease and dragonfly nymphs would increase.

D Microscopic algae would increase and dragonfly nymphs would decrease.

[END OF SECTION 1. NOW ATTEMPT THE QUESTIONS IN SECTION 2 OF YOUR QUESTION AND ANSWER BOOKLET]

X707/75/01

Biology
Section 1—Answer Grid
and Section 2

MONDAY, 9 MAY

1:00 PM – 3:00 PM

Fill in these boxes and read what is printed below.

Full name of centre

Town

Forename(s)

Surname

Number of seat

Date of birth

Day Month Year

Scottish candidate number

Total marks—80

SECTION 1—20 marks

Attempt ALL questions.

Instructions for the completion of Section 1 are given on *Page two*.

SECTION 2—60 marks

Attempt ALL questions.

Write your answers clearly in the spaces provided in this booklet. Additional space for answers and rough work is provided at the end of this booklet. If you use this space you must clearly identify the question number you are attempting. Any rough work must be written in this booklet. You should score through your rough work when you have written your final copy.

Use **blue** or **black** ink.

Before leaving the examination room you must give this booklet to the Invigilator; if you do not, you may lose all the marks for this paper.

SECTION 1—20 marks

The questions for Section 1 are contained in the question paper X707/75/02.

Read these and record your answers on the answer grid on *Page three* opposite.

Use **blue** or **black** ink. Do NOT use gel pens or pencil.

1. The answer to each question is **either** A, B, C or D. Decide what your answer is, then fill in the appropriate bubble (see sample question below).

2. There is **only one correct** answer to each question.

3. Any rough working should be done on the additional space for answers and rough work at the end of this booklet.

Sample Question

The thigh bone is called the

A humerus

B femur

C tibia

D fibula.

The correct answer is **B** — femur. The answer **B** bubble has been clearly filled in (see below).

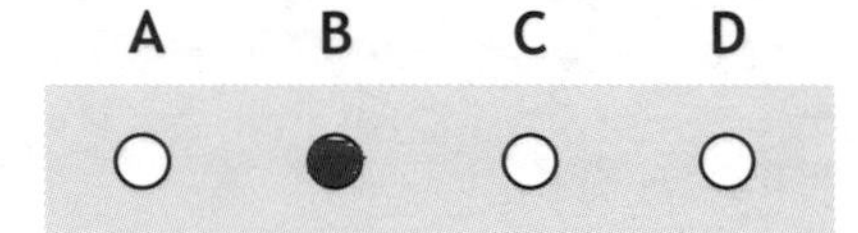

Changing an answer

If you decide to change your answer, cancel your first answer by putting a cross through it (see below) and fill in the answer you want. The answer below has been changed to **D**.

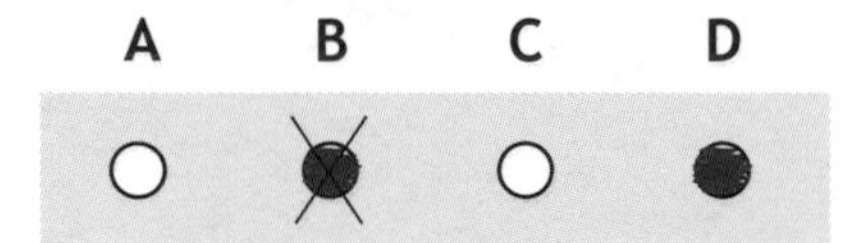

If you then decide to change back to an answer you have already scored out, put a tick (✓) to the **right** of the answer you want, as shown below:

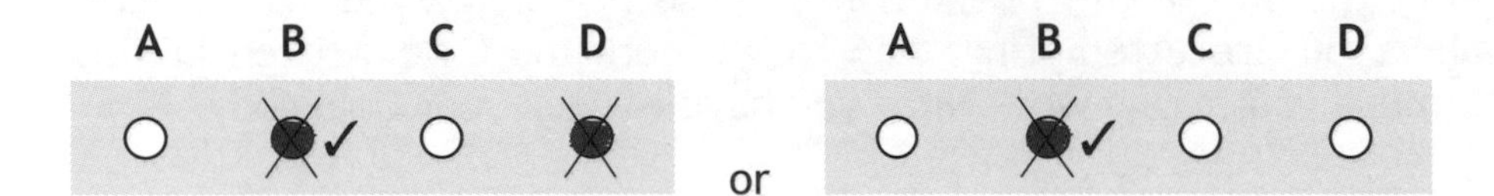

SECTION 1—Answer Grid

	A	B	C	D
1	○	○	○	○
2	○	○	○	○
3	○	○	○	○
4	○	○	○	○
5	○	○	○	○
6	○	○	○	○
7	○	○	○	○
8	○	○	○	○
9	○	○	○	○
10	○	○	○	○
11	○	○	○	○
12	○	○	○	○
13	○	○	○	○
14	○	○	○	○
15	○	○	○	○
16	○	○	○	○
17	○	○	○	○
18	○	○	○	○
19	○	○	○	○
20	○	○	○	○

[BLANK PAGE]

DO NOT WRITE ON THIS PAGE

MARKS | DO NOT WRITE IN THIS MARGIN

SECTION 2—60 marks

Attempt ALL questions

1. (a) State a feature of the cell membrane which allows the movement of only some substances into the cell. **1**

(b) Osmosis is a process which can occur across the cell membrane.

(i) Choose either the leaf cell or red blood cell by ticking (✓) one of the boxes below.

Describe the effect of osmosis on this type of cell if it was placed in pure water. **1**

Leaf cell ☐ Red blood cell ☐

Effect on the cell ______

(ii) 1 Name a process, other than osmosis, which allows molecules to pass through the cell membrane. **1**

2 Give a definition of the process chosen. **1**

[Turn over

MARKS | DO NOT WRITE IN THIS MARGIN

2. The diagram below shows how the enzyme lactase is used in the production of lactose-free milk.

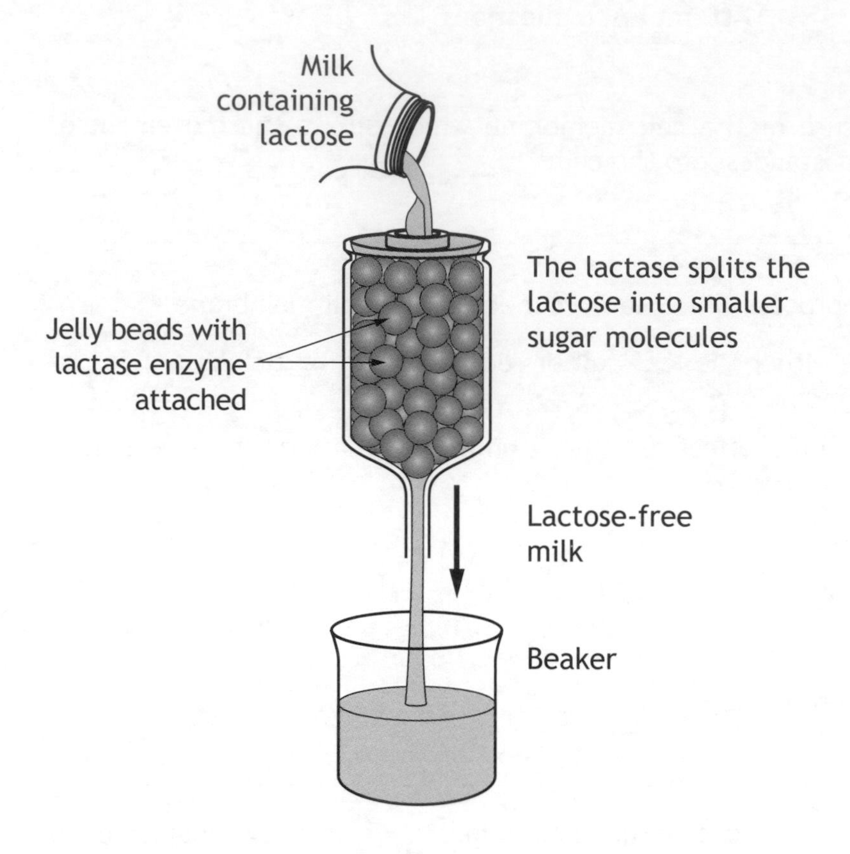

(a) (i) Underline **one** option in each of the brackets to make the following sentences correct. **2**

This process is an example of a {degradation / synthesis} reaction.

In this reaction, lactose is the {product / substrate} of lactase.

MARKS | DO NOT WRITE IN THIS MARGIN

2. (a) (continued)

(ii) A fault in the production resulted in boiling water running over the lactase enzyme.

Using your knowledge of enzymes, predict how the milk produced would differ from the expected product.

Explain your answer. **2**

Prediction ____________________

____

Explanation ____________________

____

(b) Enzymes such as lactase are biological catalysts.

Explain the role of enzymes in living cells. **1**

____

____

(c) Name the substance of which enzymes are made. **1**

____

[Turn over

MARKS | DO NOT WRITE IN THIS MARGIN

3. The diagram below represents part of the process of genetic engineering.

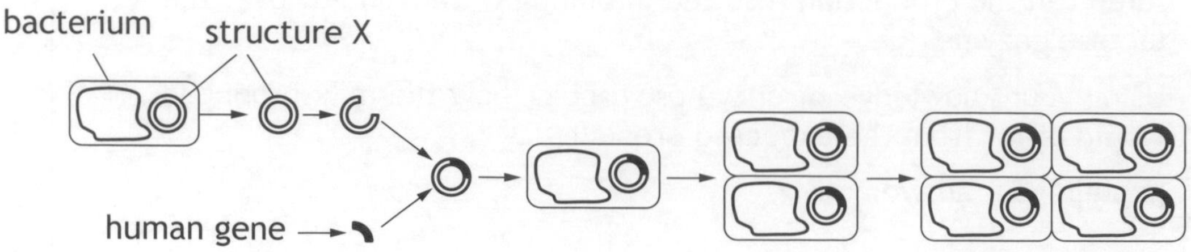

bacterium
structure X
human gene

(a) (i) Structure X is removed from the bacterium and modified during this process.

Name structure X. **1**

____

(ii) The bacteria have an initial concentration of 1000 cells/cm^3.

Each cell divides once every 30 minutes.

Calculate how long it will take for the concentration to become greater than 15 000 cells/cm^3. **1**

Space for calculation

_______________ hours

(b) The genetically modified bacteria are grown in a fermenter.

(i) Explain why the fermenter must be sterilised using aseptic techniques before it is used. **1**

____

____

(ii) The fermenter is controlled to provide optimum conditions.

Name one factor which can be controlled. **1**

____

MARKS | DO NOT WRITE IN THIS MARGIN

4. The diagram below shows muscle cells.

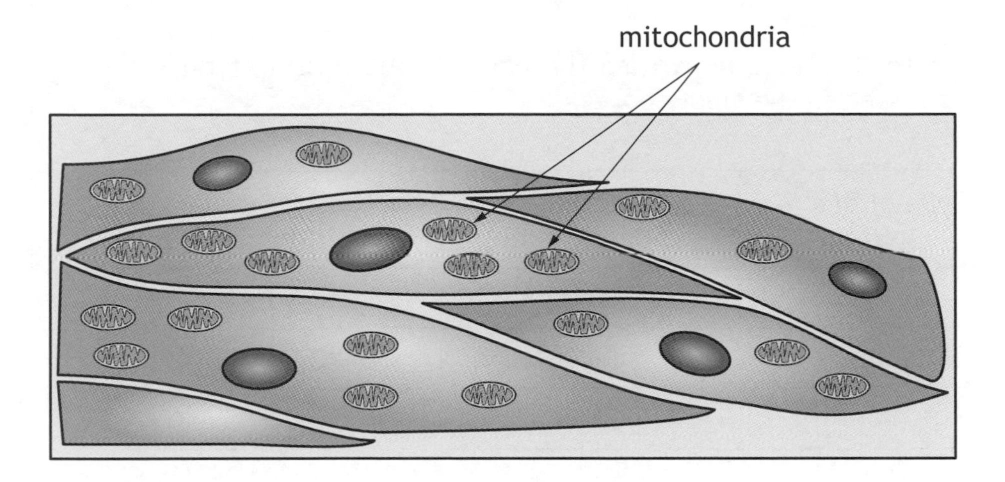

(a) (i) Explain why muscle cells require many mitochondria. **1**

(ii) Name **one** substance produced by a cell carrying out aerobic respiration. **1**

(b) A muscle cell will carry out fermentation when oxygen is not available.

Describe the fermentation pathway in muscle cells. **3**

MARKS DO NOT WRITE IN THIS MARGIN

5. The table below gives information about features of three different types of blood vessel.

(a) (i) Complete the table by writing the name of the missing types of blood vessels in the empty boxes. 2

Type of blood vessel	*Diameter of central channel* (mm)	*Thickness of vessel wall* (mm)
	30·0	1·5
Capillary	0·006	0·001
	25·0	2·0

(ii) Of all the blood vessels, capillaries are best adapted for gas exchange.

Using the information in the table, give a reason for this. 1

(b) The heart is a muscle which pumps blood around the body and requires its own blood supply.

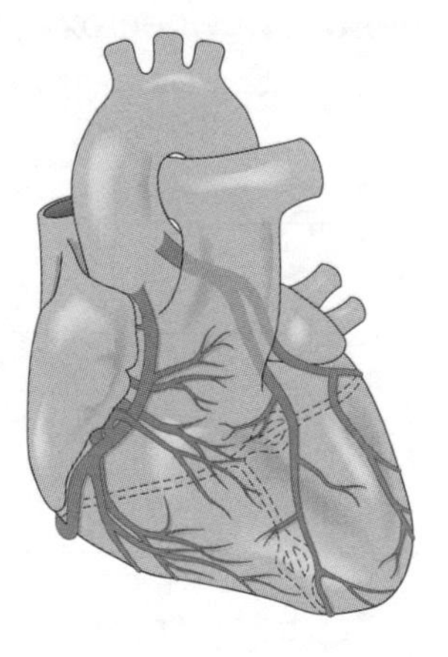

Name the blood vessel which supplies the heart muscle with blood. 1

MARKS | DO NOT WRITE IN THIS MARGIN

6. The following diagram represents part of a family tree showing the inheritance of hitchhiker's thumb, where the thumb can bend back as shown below.

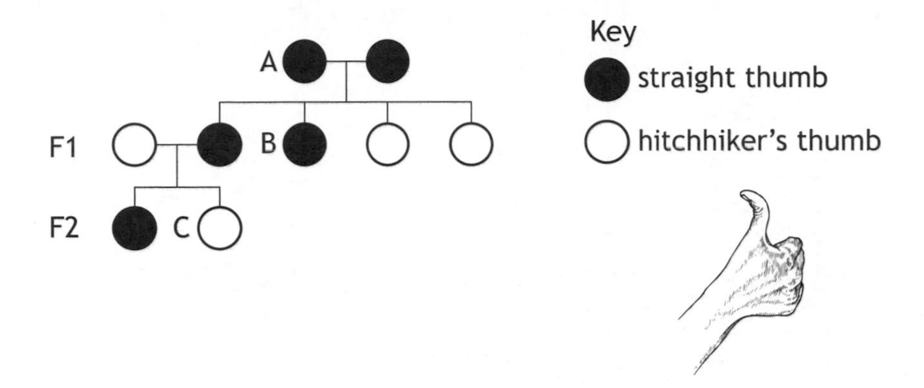

(a) Complete the table below for individuals A and C. **2**

Individual	*Possible Genotype(s)*	*Phenotype*
A		straight thumb
B	TT or Tt	straight thumb
C	tt	

(b) In a survey of 90 students it was found that 25 of them had hitchhiker's thumb.

(i) Calculate the number of students with straight thumb to hitchhiker's thumb as a simple, whole number ratio. **1**

Space for calculation

__________ : __________

straight thumb   hitchhiker's thumb

(ii) The predicted ratio was 3 straight thumb : 1 hitchhiker's thumb.

Explain why the predicted ratio was different to the actual ratio. **1**

____

____

[Turn over

MARKS DO NOT WRITE IN THIS MARGIN

7. (a) The rate of transpiration in plants can be measured using the apparatus shown below.

As the plant transpires, coloured water is drawn up the glass tube and its volume measured, over a set period of time, to give the rate of transpiration.

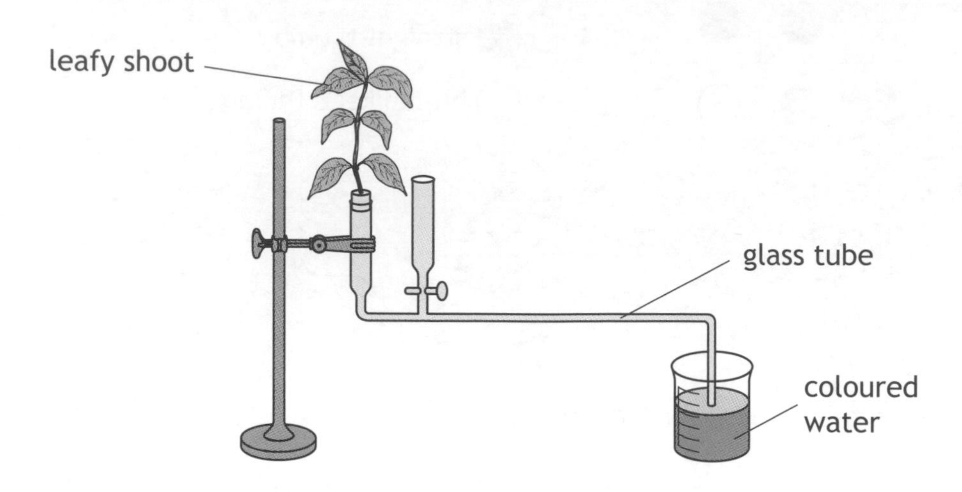

Changes in the surrounding environment can have an effect on the rate of transpiration.

(i) Select **one** of the environmental changes listed below by  it.

increase in humidity	increase in temperature	increase in wind speed

State the effect of this change on the rate of transpiration. 1

(ii) Choose any of the environmental changes listed above and describe an addition to the apparatus shown, which would allow an investigation into its effect. 1

Environmental change ______

Description of addition ______

MARKS DO NOT WRITE IN THIS MARGIN

7. (continued)

(b) The graph below shows transpiration rates of two plants, P and Q.

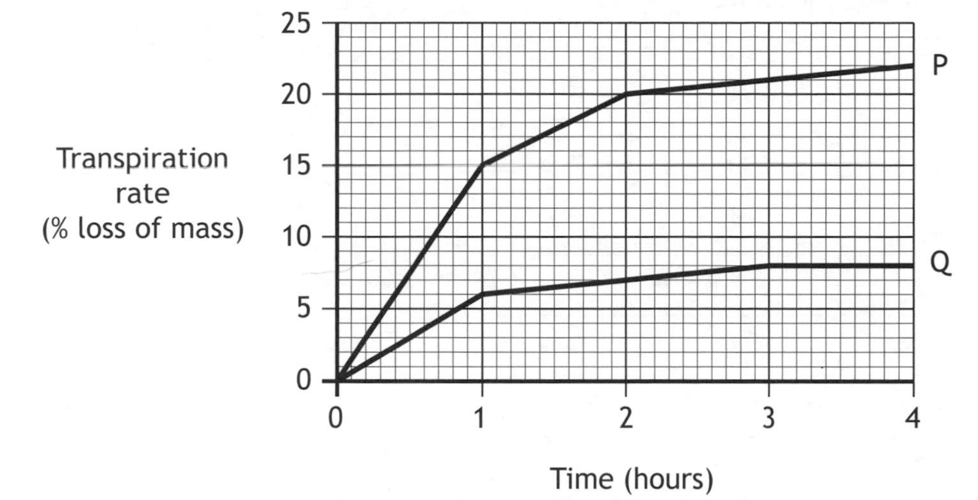

(i) With reference to the number of stomata, suggest a reason for the different transpiration rates of plants P and Q. **1**

(ii) Name the type of cells which control the opening and closing of stomata. **1**

[Turn over

MARKS DO NOT WRITE IN THIS MARGIN

8. Nutritional information helps people make an informed choice about the food they eat.

Table 1 – Label from a bar of chocolate

Nutritional information	*per 100 g*	*per bar*	*% RI**
Energy (kJ)	2251	630	7·5
Sugar	65 g	18 g	15·6
Protein	10 g	2·8 g	3
Total fat	25 g	7 g	10
Saturated fat	20 g	5·6 g	28
Salt	0·4g	0·1 g	1·7

*RI = Reference Intake (formerly "guideline daily amount")

Table 2 – Guidelines on salt content

Salt category	*Salt content (g/100 g)*
High	More than 1·5
Medium	0·3 to 1·5
Low	Less than 0·3

(a) Using information from **Table 1 and Table 2**, identify the salt category to which this chocolate bar belongs. **1**

____

MARKS | DO NOT WRITE IN THIS MARGIN

8. (continued)

(b) Use the information in **Table 1** to complete the pie chart below to show the composition of protein, sugar and total fat in 100 g of the chocolate. **2**

(An additional pie chart, if required, can be found on *Page twenty-six*)

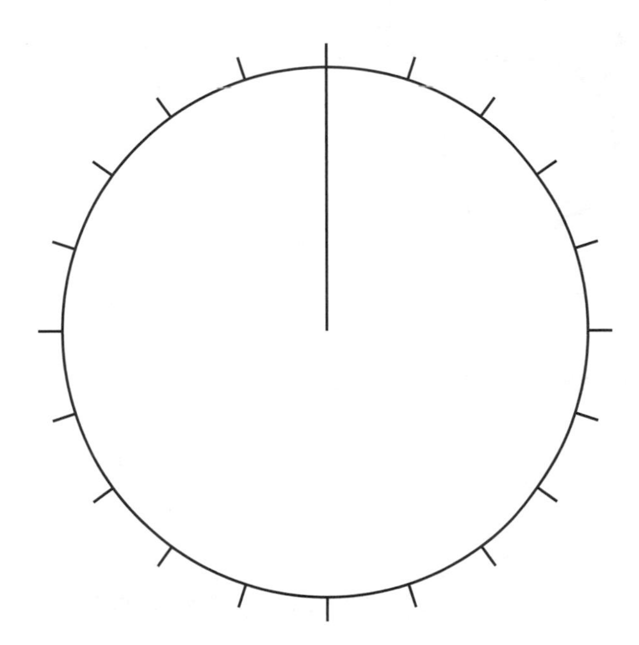

(c) (i) As shown in Table 1, saturated fat makes up part of the total fat in this chocolate bar.

Calculate the percentage of total fat which is saturated. **1**

Space for calculation

____________ %

(ii) One bar of this chocolate contains 630 kilojoules which is 7·5% of the reference intake (RI).

Calculate the total number of kilojoules which should be consumed daily. **1**

Space for calculation

____________ kilojoules

MARKS | DO NOT WRITE IN THIS MARGIN

9. (a) The diagram below represents a hormone binding to a cell within its target tissue.

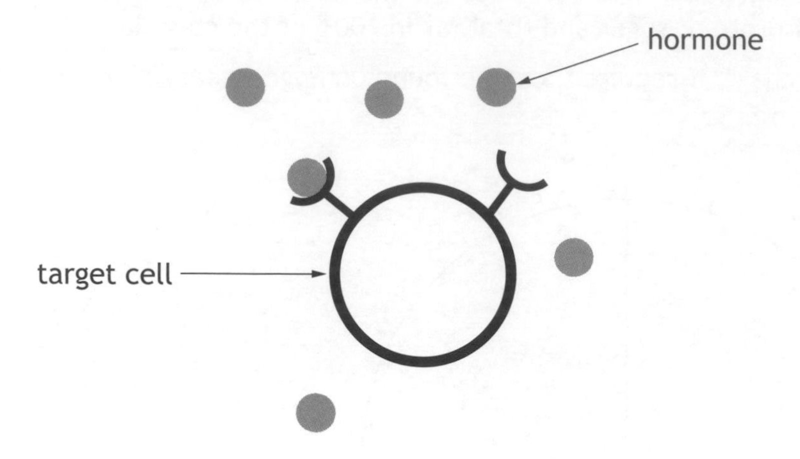

Explain why only the target cells are affected by this hormone. 1

____

____

(b) Name the type of gland that releases hormones into the bloodstream. 1

____

(c) Blood glucose levels are controlled by two hormones.

<u>Underline</u> one option in the bracket to make the following sentence correct. 1

A decrease in blood glucose levels is detected by the pancreas

and this causes an increase in the release of {glycogen / insulin / glucagon}

into the blood stream.

MARKS | DO NOT WRITE IN THIS MARGIN

10. A food chain from a river is shown below.

algae ⟶ water flea ⟶ stickleback ⟶ perch

Using the information in the food chain, answer the following questions.

(a) (i) Identify an organism which is **both** predator and prey. **1**

____

(ii) Pesticides are known to run off from the land into rivers and enter the food chains.

Name the organism which would accumulate the greatest concentration of pesticides in its body over a period of time. **1**

____

(b) State **one** way in which energy may be lost between stages in a food chain. **1**

____

[Turn over

MARKS | DO NOT WRITE IN THIS MARGIN

11. (a) In an investigation, students estimated the population and biomass of some organisms found on part of a rocky shore.

The table below shows the results.

Organism	*Population*	*Average mass of one organism* (g)	*Biomass of population* (g)
Seaweed	220	500	110 000
Limpet	1 100		33 000
Crab	100	90	9 000
Gull	5	700	3 500

(i) Complete the table to show the average mass of one limpet. **1**

Space for calculation

(ii) The total mass of living material decreases at each level in the food chain. This can be shown as a pyramid of biomass.

Complete the diagram below by entering the names of the organisms from the table into the appropriate section. **1**

(An additional diagram, if required, can be found on *page twenty-six*)

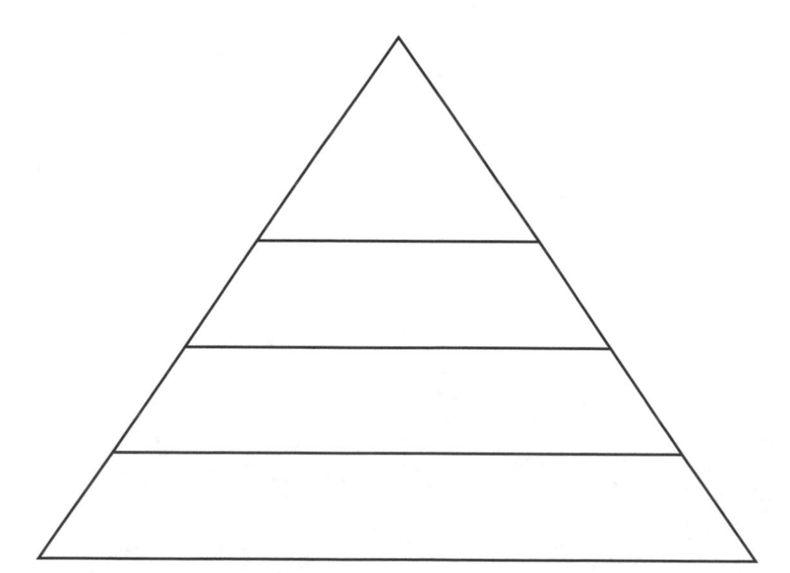

MARKS DO NOT WRITE IN THIS MARGIN

11. (continued)

(b) During the investigation the students found four different species of periwinkles at different positions on the rocky shore.

The highest position that the sea water reaches on the shore is called the high tide level.

The bars in the table below represent the positions on the shore where each species of periwinkle was found.

	Species of periwinkle			
Position on shore	*Small*	*Edible*	*Rough*	*Flat*
High tide level ↓ Low tide level				

(i) State which species of periwinkle is least likely to compete with the small periwinkle.

Explain your answer. **1**

Species ______________________

Explanation ______________________

____

(ii) Using the information given, explain why the competition between these periwinkles is described as interspecific. **1**

____

____

[Turn over

MARKS | DO NOT WRITE IN THIS MARGIN

12. A group of students carried out a five year investigation into plant growth in an area of abandoned farmland.

They sampled the area using quadrats.

The results are shown in the table below.

	Average abundance of each plant		
Year	*Meadow grass*	*Ragwort*	*Pink campion*
2011	8	15	9
2012	16	14	7
2013	24	12	4
2014	25	8	2
2015	25	5	1

(a) (i) Calculate the average decrease per year in the abundance of ragwort over the five-year period. **1**

Space for calculation

____

(ii) **Use information from the table** to suggest why the ragwort abundance decreased over the five-year period. **1**

____

____

____

(b) The students also sampled invertebrates such as beetles and spiders.

Name a sampling technique they could have used and describe a possible source of error with this technique. **2**

Sampling technique __________

Source of error __________

____

MARKS DO NOT WRITE IN THIS MARGIN

12. (continued)

(c) The following table gives information about some of the flowering plants found in the area.

Plant	*Height range* (cm)	*Flower colour*	*Flowering period* (months)
Pink campion	30-90	pink	6
Ragwort	30-200	yellow	6
Meadow grass	30-70	green	3
Buttercup	5-90	yellow	5

Using the information in the table, complete the three boxes in the paired statement key below. **3**

1. Flower colour is yellow — go to 2

   Flower colour is not yellow — [ ]

2. Height of plant can be over 100 cm — Ragwort

   Height of plant is under 100 cm — [ ]

3. Flowering period lasts only 3 months — Meadow Grass

   Flowering period is longer than 3 months — [ ]

[Turn over

MARKS | DO NOT WRITE IN THIS MARGIN

13. The diagrams below show the light and dark varieties of a moth which can be found in woodland areas. These moths rest on the bark of trees during the day and can be eaten by birds. Normally the bark of trees in the woodland is light coloured. However in industrial areas, pollutants cause the tree bark to darken.

Woodland area

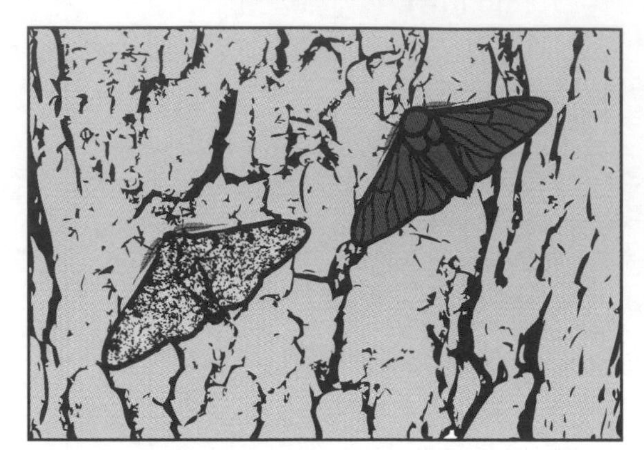

Industrial area

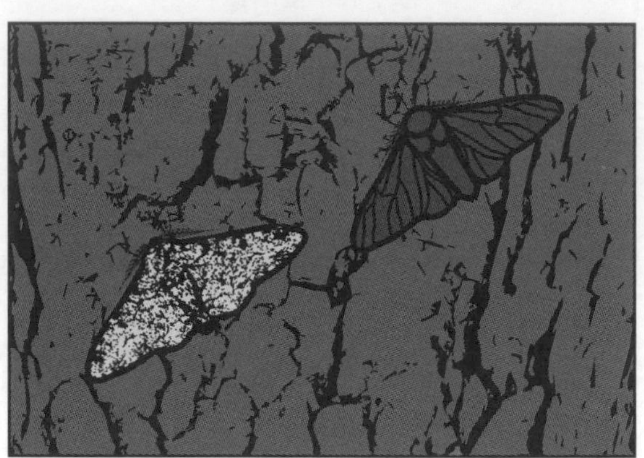

(a) The dark variety of the moth is the result of a random change in the genetic information.

State the term used to describe this change. **1**

____

(b) An investigation into the population of these moths in a woodland was carried out. The moths were captured, marked and released. 24 hours later the moths were recaptured.

The results are shown in the following table.

Variety of moth	*Number of moths marked and released*	*Number of marked moths recaptured*	*Marked moths recaptured* (%)
Light	480	264	55
Dark	520	208	40

(i) Suggest a reason why the number of the marked moths recaptured was worked out as a percentage. **1**

____

____

MARKS | DO NOT WRITE IN THIS MARGIN

13. (b) (continued)

(ii) The woodland was in a non-industrial area.

Explain why the percentage of light moths recaptured was higher than dark moths. **1**

(iii) Name the process which results in the better adapted variety of moth being more likely to survive and reproduce. **1**

[Turn over

MARKS | DO NOT WRITE IN THIS MARGIN

14. Red spider mites are a common pest which destroy tomato plants. Some of the mites are resistant to chemical pesticides.

Tomato growers aimed to investigate whether a predator would reduce the spider mite numbers in their greenhouses. Two identical greenhouses were used and the predator was released into only one greenhouse.

The results are shown in the graph below.

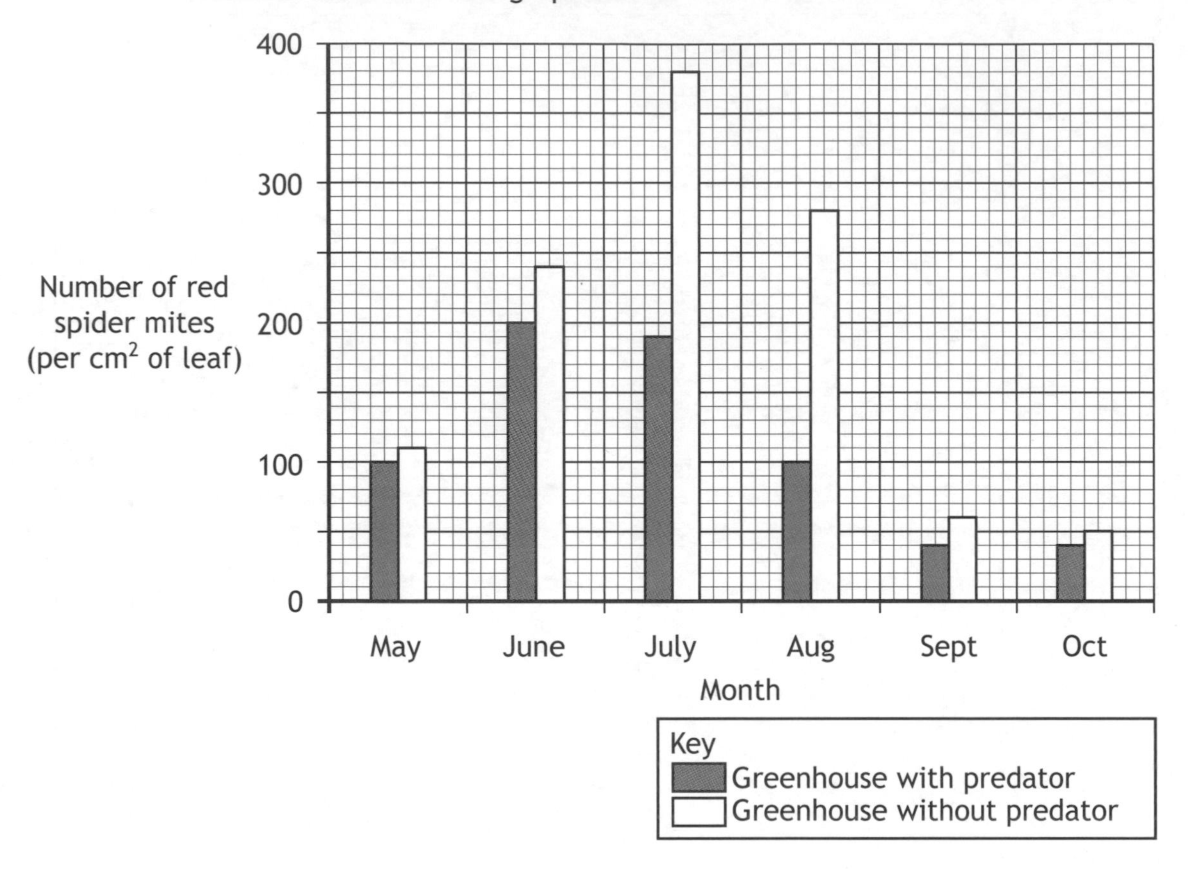

(a) (i) With reference to the aim of this investigation, give the conclusion that the tomato growers would have drawn from these results. **1**

____

____

MARKS DO NOT WRITE IN THIS MARGIN

14. (a) (continued)

(ii) The greenhouse containing tomato plants without predators was included as a control experiment.

State the purpose of the control in this investigation. **1**

(b) State the term which describes the use of a predator as an alternative to pesticides. **1**

[END OF QUESTION PAPER]

MARKS DO NOT WRITE IN THIS MARGIN

ADDITIONAL SPACE FOR ANSWERS AND ROUGH WORK

ADDITIONAL PIE CHART FOR QUESTION 8(b)

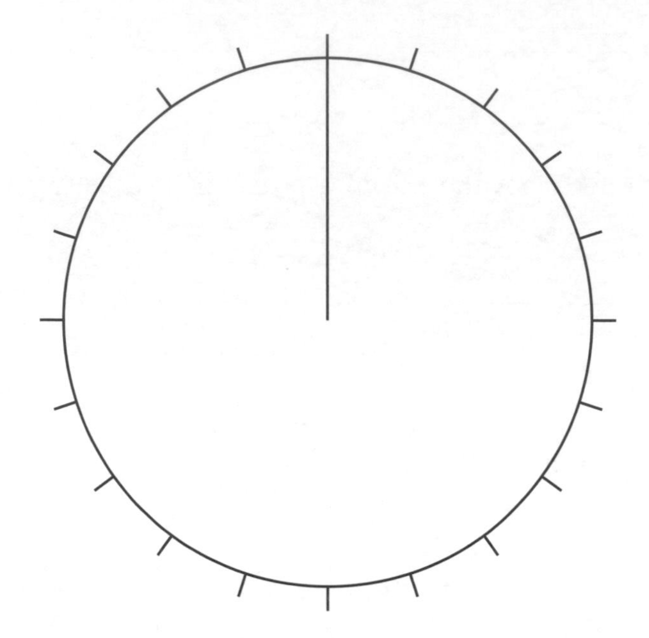

ADDITIONAL DIAGRAM FOR QUESTION 11(a) (ii)

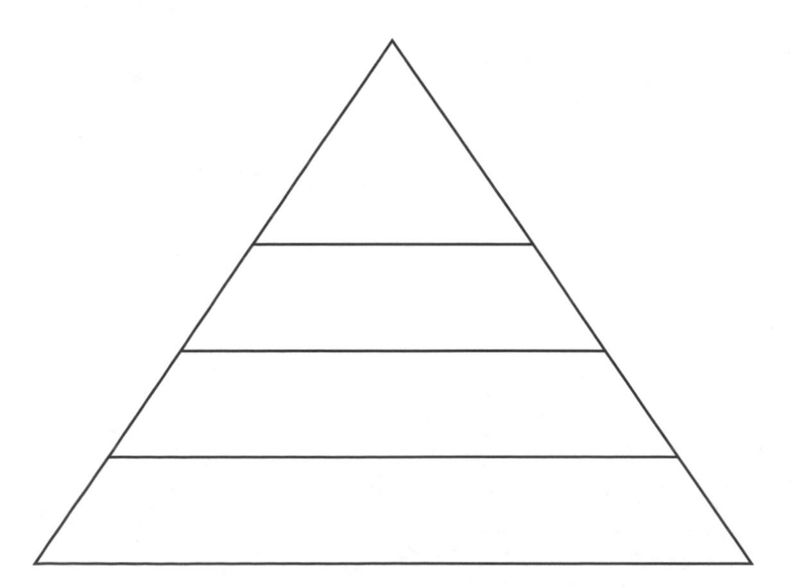

MARKS DO NOT WRITE IN THIS MARGIN

ADDITIONAL SPACE FOR ANSWERS AND ROUGH WORK

MARKS | DO NOT WRITE IN THIS MARGIN

ADDITIONAL SPACE FOR ANSWERS AND ROUGH WORK

NATIONAL 5

Answers

ANSWERS FOR

SQA NATIONAL 5 BIOLOGY 2016

NATIONAL 5 BIOLOGY 2014

SECTION 1

Question	Response
1.	B
2.	D
3.	B
4.	C
5.	D
6.	C
7.	A
8.	C
9.	A
10.	A
11.	C
12.	A
13.	D
14.	B
15.	C
16.	C
17.	D
18.	A
19.	B
20.	D

SECTION 2

1. (a) (i) Plant, bacterial and fungal

(ii) Bacteria(l)

(iii) Ribosome – (site of/involved in) protein synthesis

Mitochondria – (site of/involved in) energy or ATP production/aerobic respiration

(b) Y axis scale and label, including units

Bars correctly plotted

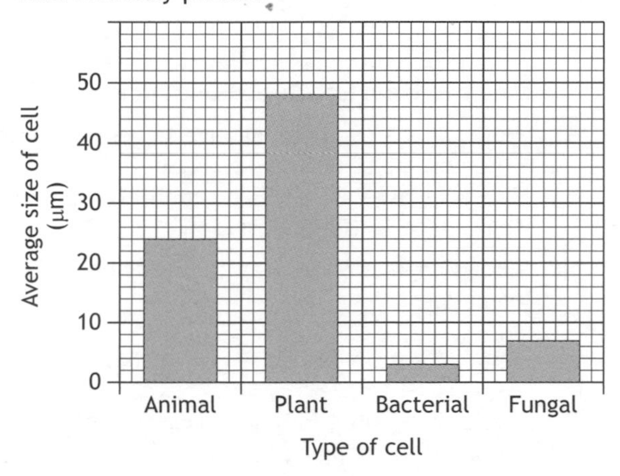

2. (a) Osmosis

(b) Water moves into the (model) cell/bag/salt solution 1 mark

From a high water concentration to a low water concentration/down a concentration gradient 1 mark

OR

Alternative answer for 2 marks:

Water moves from a high water concentration outside to a low water concentration inside the (model) cell/bag/ salt solution

(c) 0.9

(d) Description of concentration change – must be a smaller concentration gradient than shown/lower temperature/wider capillary tube/seal not tight/ less water in the beaker/bag not fully submerged

3. (a) (i) Hydrogen peroxide

(ii) Numbers (in each group) different

OR

Overall numbers used too small

(iii) If they have a low level of catalase/only use sheep with low levels of catalase/don't use sheep with high levels of catalase

(b) (Activity is) decreased/slows down reaction

4. A villus has:

A thin wall

A large surface area

A good blood supply/many capillaries

- There are a large number of villi
- So this also increases surface area/creates a large surface area

  OR

  makes absorption/diffusion fast(er)/more

5. (a)

Statement	*Stage 1*	*Stage 2*
Carbon dioxide required		✓
Light energy required	✓	
Water required	✓	
Sugar produced		✓
ATP + Hydrogen required		✓
Oxygen produced	✓	

1 mark for each correct column

(b) Photosynthesis is controlled by enzymes/enzymes are needed

(At high temperatures) enzymes are denatured/do not work.

(c) Light Intensity, Temperature } either order

6. (a) C A B F E D

(b) (Sperm)

F Reproduction/fertilisation (or correct description)

E Tail to swim/mitochondria for energy/haploid to allow fusion at fertilisation producing a diploid zygote

OR

Egg cell

F Reproduction/fertilisation (or correct description)
E Yolky cytoplasm/large cell to provide food or haploid to allow fusion at fertilisation producing a diploid zygote

OR

Red blood cell

F Carry oxygen
E Contain haemoglobin/carries oxygen as oxyhaemoglobin
OR
large surface area/ biconcave/no nucleus to transport more oxygen
OR
small/flexible to go through capillaries

(c) Unspecialised/undifferentiated/not specialised

(d)

Growth of new skin	✓
Transmission of nerve impulses	
Muscle contraction	
Repair of broken bones	✓
Production of insulin	

7. (a) Aerobic respiration

(b) (i) Glucose
(ii) ADP + Phosphate/Pi/ $PO_4 \longrightarrow$ ATP

(c) (i) Sprinter
Highest lactic acid produced when oxygen is not used to release energy
OR
Highest percentage light tissue
OR
Highest fermentation
OR
Highest percentage of cells that do not use oxygen.
(ii) Swimmer

8. (a) (i) A = low B = high
C = glucagon D = insulin
(ii) Organ X = pancreas
Organ Y = liver

(b) *Any two features from:*
- Made of protein
- Chemical messengers
- Specific for some (target) tissues
- Shaped to fit receptors
- Released or produced by endocrine glands/ system
- Carried in blood
- Can have a long term effect

9. (a) Genotypes: BB and bb
F_1 phenotype: black (coat)

(b) (i) More than one/several genes control one/a characteristic
(ii) Continuous

10. (a) (i) • (The general trend is) as the distance increases, numbers/population/lugworms increases up to 12 metres
• After that, numbers/population remains steady/stays the same
(ii) 3

(b) (i)

Predator	*Type of Competition*	
	Intraspecific	*Interspecific*
rex sole and curlew		✓
curlew and curlew	✓	
rex sole and dover sole		✓

(ii) 16.5 kJ

11. (a) Named abiotic factor, eg moisture, pH, light intensity
Description of method, including instrument name, eg "Stick the probe of the pH meter in the soil"

(b) (i) Left traps too long/
Traps too high above soil/
Traps not camouflaged/
Traps too shallow
(ii)

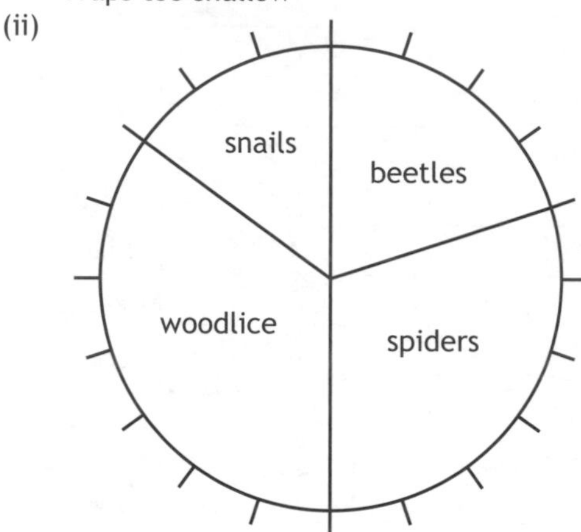

1 mark for appropriate sized sections
1 mark for labels

(c) Able to fly/flew away

12. (a)
- Initial population is separated/split (or idea of this)
- (Different) mutations occur in each subpopulation/group (need indication that it is more than the original one population)
- Some mutations are advantageous
- Natural selection occurs
  OR
  Selection pressures are different in each group
  OR
  Advantageous mutations are selected for
- Subpopulations/groups are no longer able to interbreed to produce fertile offspring

Any two from last three bullet points

(b) Mutation – a (random) change to genetic material/ chromosome structure or number/bases in DNA
Species – organisms which can interbreed/reproduce to produce fertile offspring

(c) Allows population to adapt to changing environmental conditions
OR
suitable example of coping with change
OR
makes it possible for population to evolve in response to changing conditions

NATIONAL 5 BIOLOGY 2015

SECTION 1

Question	Response
1.	D
2.	A
3.	C
4.	D
5.	B
6.	A
7.	D
8.	D
9.	C
10.	B
11.	A
12.	C
13.	D
14.	B
15.	B
16.	B
17.	C
18.	C
19.	C
20.	B

SECTION 2

1. (a) (i) 4
   (ii) 4
   (b) Meristem(s)/shoot tip/root tip
2. (a) (i) +25
   (ii) To remove excess/surface water/liquid/solution.
   OR
   So water/liquid/solution doesn't affect the results or alter the mass/weight.
   (iii) Beaker A
   Water entered **(1)** (the egg) from a high water concentration (outside) to a low water concentration (inside)/ down a concentration gradient. **(1)**
   OR
   Beaker B
   Water left/leaves **(1)** (the egg) from a high water concentration (inside) to a low water concentration (outside)/down a concentration gradient. **(1)**
   (b) Passive transport doesn't require energy/ATP, but active transport does.
   OR
   Passive transport moves down a concentration gradient/from high to low, but active transport goes up/against a concentration gradient/from low to high.
3. (a) TAC GCT ACG CGA CAG
   (b) (i) Protein
   (ii) Molecule P: mRNA/messenger RNA **(1)**
   Description: The order/sequence of bases (determines the order/sequence of amino acids) **(1)**
   (iii) Nucleus
4. (a) Name of first stage: light reactions **(1)**
   Diffuses out of the leaf: oxygen **(1)**
   Two products used in second stage: hydrogen and ATP **(Both = 1)**
   (b) Forms sugar/glucose/starch **(1)**
   ATP provides energy
   OR
   Hydrogen combines/reacts/joins with CO_2 **(1)**
5. (a) Aerobic respiration
   (b) 3
   (c) (Sperm) require more energy/ATP
   AND
   as they move (more)/are (more) active/to swim.
6. (a) Electrical impulse/electrical message/electrical signal
   (b) J = Sensory (neuron) – carries/sends message/impulse/signal from sense organ to relay neuron/CNS/spinal cord.
   K = Motor (neuron) – carries/sends message/impulse/signal from CNS/relay neuron/spinal cord to muscle/organ/effector.
   L = Relay (neuron) – carries/sends message/impulse/signal from sensory to motor neuron/within CNS.
   (c) 0.01
7. (a) (i) Jon is heterozygous/Hh/has both alleles/both forms of the genes
   AND
   is hearing. **(Both = 1)**
   (ii)

Person	*Genotype*	*Phenotype*
Paul	hh	non-hearing
Lyall	Hh	hearing

(1 mark per column)

   (iii) 3 in 4/75%
   (b) Polygenic
8. (a) (i) Both scale and axis label completed correctly **(1)**
   Points plotted correctly and joined **(1)**
   (ii) 1300
   (b) Water moves into/enters/is absorbed by root hairs by osmosis/diffusion. **(1)**
   (Water) travels upwards in the xylem. **(1)**
   (Water) travels to the stomata/pores and evaporates/transpires/diffuses out. **(1)**
9. (a) (i) Alveolus/alveoli/air sac
   (ii) Large surface area
   Thin walls/walls are one cell thick
   Good/rich blood supply/dense capillary network
   Moist
   (Any two = 1 mark each)

(b) Dirt/dust/microorganisms are trapped in the mucus. **(1)**
Cilia move these up and away from the lungs. **(1)**

10. (a)

Stage	*Number*
	3, 4 or both
	8

(1 mark per row)

(b) Soil/root nodules

(c) R = nitrate(s) **(1)**
Needed to make protein **(1)**

11. (a) (i) As the (number of) bacteria increases, the oxygen (level in the water) decreases.
OR
As the (number of) bacteria decreases, the oxygen (level in the water) increases.

(ii) 2

(b) (i) Mayfly nymphs/stonefly nymphs/caddis fly larvae

(ii) (The pollution/sewage results in) fewer/less types (of organism/animals).
OR
(Pollution) decreases biodiversity.
OR
'They would decrease' (as this refers to the types of organisms).

(c) Organisms which, by their presence/absence, show level of pollution / environmental quality.

12. (a) 36

(b) Continuous

(c) Too few leaves/taken/sampled.
OR
More than 5 leaves should be measured.
OR
Only five leaves were taken.
OR
Too small a sample.

(d) To make sure the leaves belong to the same species/type of ivy.
OR
Different plants could be affected by different factors/growing conditions.
OR
So there is only one variable.
OR
To limit/reduce the number of variables.

(e) Light intensity/temperature/wind speed/humidity (or moisture in the air)

13. (a) (i) 65

(ii) To prove that the mutation was causing the effect/high bone density (or equivalent description).

(b) Radiation or example
OR
Chemicals or example
OR
High temperature

(c) • (Allows the species) to evolve/adapt
• in response to changing environmental conditions.
(Both = 1)

NATIONAL 5 BIOLOGY 2016

SECTION 1

Question	Response
1.	B
2.	A
3.	B
4.	D
5.	C
6.	C
7.	A
8.	C
9.	A
10.	B
11.	B
12.	C
13.	C
14.	B
15.	D
16.	B
17.	C
18.	A
19.	D
20.	D

SECTION 2

1. (a) Selectively permeable/semi-permeable/(contains) proteins/(phospho)lipids/protein channels/protein carriers

(b) (i) Leaf:
• cell swells/becomes turgid (or suitable description of turgid)
Red blood cell:
• cell swells/bursts/may burst

(ii) 1. Diffusion/active transport
2. Definition:
Diffusion – Movement of molecules/particles from a high to a low concentration
OR
down the concentration gradient
Active Transport – Movement of molecules/ions from a low to a high concentration
OR
against/up the concentration gradient

2. (a) (i) Degradation **(1)**
Substrate **(1)**

(ii) **Prediction** - (All or some) lactose would not be removed from the milk/milk would contain lactose/it would not be lactose free **(1)**
Explanation - Enzyme/lactase denatured
OR
Enzyme/active site has changed shape/description of change of shape **(1)**

(b) Speed up (chemical/biological/biochemical) reactions/allow reactions to occur at lower temperatures/lower the activation energy

(c) Protein/amino acids

3. (a) (i) Plasmid
(ii) 2

(b) (i) To ensure there are no other microbes/bacteria (or equivalent) present
OR
To prevent/stop contamination/cross-contamination/growth of other cultures
(ii) Temperature/pH/0_2 or $C0_2$ concentration/ nutrient or food levels

4. (a) (i) Requires/uses/needs a lot of energy/ATP
AND
For movement/contraction
(ii) Carbon dioxide/Water/(38) ATP

(b) Glucose converted/broken down to pyruvate/ pyruvic acid **(1)**
Pyruvate/pyruvic acid converted to lactic acid **(1)**
(2) ATP produced **(1)**

5. (a) (i)

Type of blood vessel	
vein	**(1)**
artery	**(1)**

(ii) They have thinnest/thinner wall(s)

(b) Coronary artery/arteries

6. (a)

Individual	*Possible Genotype(s)*	*Phenotype*	
A	Tt		**(1)**
B			
C		Hitchhiker's (thumb)	**(1)**

(b) (i) 13:5
(ii) Fertilisation is a random process
OR
Numbers in sample too small

7. (a) (i) increase in humidity - decreases

increase in temperature - increases

increase in wind speed - increases

(ii) **humidity**
put the apparatus in a (transparent) bag/ container
temperature
put a heater beside it/put in a water bath at a higher temperature
wind speed
use a fan/hairdryer on cool setting beside the apparatus

(b) (i) P has a greater number of stomata/Q has fewer stomata
(ii) Guard (cells)

8. (a) Medium (salt)

(b)

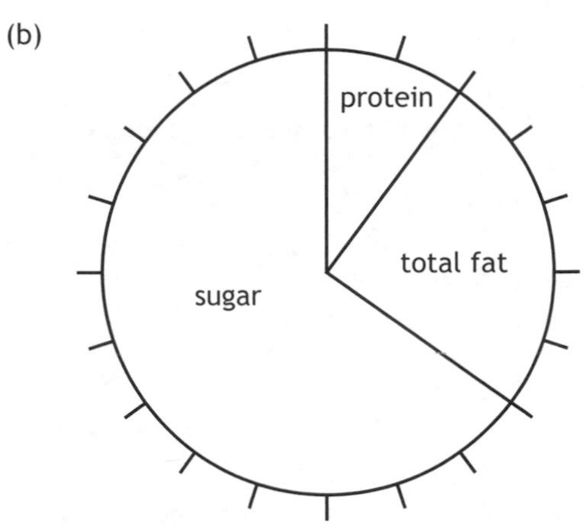

(1 mark for divisions, 1 mark for labels)

(c) (i) 80
(ii) 8400

9. (a) They have receptors/receptor proteins
AND
these are specific/match this hormone

(b) Endocrine

(c) Glucagon

10. (a) (i) Stickleback
(ii) Perch

(b) Heat/movement/undigested material/faeces/ excrement/fur/bones/hair

11. (a) (i) 30
(ii)

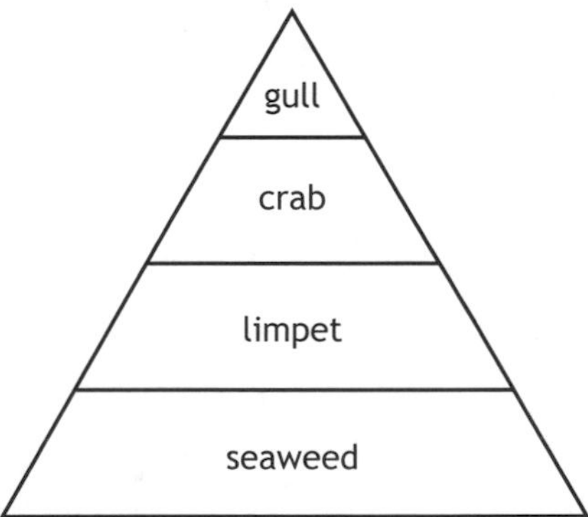

(b) (i) Flat (periwinkle)
Don't live on/occupy the same position on the shore/live on different/separate parts of the shore/small live at high tide and flat live at low tide/one lives at low tide and one at high tide
(ii) They are different species/they are not the same species/more than one species are competing

12. (a) (i) 2
(ii) **Increased** competition from Meadow grass or appropriate description of the increased competition eg less space for Ragwort to grow

(b) Sampling Technique: Pitfall trap **(1)**
Source of error:
Traps left too long/not checked regularly
Too high above soil surface/too low below soil surface/not level with soil surface
Not camouflaged
Too shallow
No drainage holes **(any one - 1)**

(c) Go to 3 **(1)**
Buttercup **(1)**
Pink campion **(1)**

13. (a) Mutation

(b) (i) Different numbers released/marked/captured
OR
To compare results

(ii) Fewer were eaten (by predators/birds)/better camouflaged so not eaten/camouflaged from predators/birds less likely to be eaten/seen by predators or birds/more dark moths eaten by predators or birds

(Answer must have reference to 'being eaten' or 'predators/birds')

(iii) Natural selection/survival of the fittest

14. (a) (i) When predators are present (the number of red spider) mites decrease/there are more (red spider) mites when there is no predator or converse

(ii) To allow it to be compared to the one with the predator/to compare the number of (red spider) mites with and without the predator/to show any difference is due to the predator

(b) Biological control

Acknowledgements

Permission has been sought from all relevant copyright holders and Hodder Gibson is grateful for the use of the following:

Image © Darren Baker/Shutterstock.com (2014 Section 2 page 6);
Image © Blamb/Shutterstock.com (2014 Section 2 page 11);
Image © Ysbrand Cosijn/Shutterstock.com (2014 Section 2 page 19);
Image © Alena Brozova/Shutterstock.com (2015 Section 2 page 15);
Image © Reika/Shutterstock.com (2015 Section 2 page 20);
Image © Viktor Gladkov/Shutterstock.com (2015 Section 2 page 22);
Image © Pan Stock/Shutterstock.com (2016 Section 2 page 19);
Image © HHelene/Shutterstock.com (2016 Section 2 page 24).